HUMAN IMPACT ON GANGA RIVER ECOSYSTEM

गङ्गा गङ्गेति यो ब्रूयात् योजनानां शतैरपि ।
मुच्यते सर्वपापेभ्यो विष्णुलोकं स गच्छति ॥

—स्नानविधिप्रयोग
भारतीय संस्कृतिकोश
खंड २, पृष्ठ ६७४ (गङ्गा)

Even if separated by hundreds of miles, anyone reciting the name of Ganga repeatedly, would be absolved of all the sins, and would attain 'Moksha'.

—Snanavidhiprayoga
from Bhartiya Sanskritikosha
Vol. II, p. 674. (Ganga)

Editor

Pt. Mahadevshastri Joshi
Bharatiya Sanskritikosha Mandal, Pune.

HUMAN IMPACT ON GANGA RIVER ECOSYSTEM

An Assessment

S.B. CHAPHEKAR
G.N. MHATRE

CONCEPT PUBLISHING COMPANY, NEW DELHI

First Published 1986

Sharad Bhalchandra **Chaphekar** (*b*. 1935-
Girish Nanabhai **Mhatre** (*b*. 1955-)

Published by

Naurang Rai
Concept Publishing Company
H-13, Bali Nagar
NEW DELHI-110015 (India)

Printed in India by

Lohia Composing Agency
at Sunil Printers
CB-1067, 75/1, Ajay Palace, Naraina, Ring Road
NEW DELHI-110028

Dedicated
to
the memory of
late Prof. Barun C. Haldar

Project Personnel

Late Prof. B.C. Haldar	Project Director and Principal Investigator
Dr. S.B. Chaphekar	Principal Investigator and Project Director (from October, 1981)
Dr. P.H. Umadikar	Principal Investigator
Dr. G.N. Mhatre	Scientific Officer
Miss S.B. Sawant	Administrative Secretary (from 19-7-1981 to 31-8-1981)
Miss M.M. Amdekar	Administrative Secretary (from 1-9-1981)

Foreword

THE history of India is interwined with its river systems, and out of the 14 such systems, the Ganga stands apart for reasons more than one. It has been aptly called the "River of India". The Ganga basin drains eight states : Himachal Pradesh, Punjab, Haryana, Uttar Pradesh, Rajasthan, Madhya Pradesh, Bihar, West Bengal and Union Territory of Delhi. Pandit Jawaharlal Nehru has remarked, "The story of the Ganga from source to the sea, from old times to new, is the story of India's civilisation and culture". The river system is both holy and a lifeline for millions of people. Once a symbol of purity, it is today subjected to indiscriminate overuse and abuse on account of biological and chemical pollution and heavy silt load. The biological pollution ranges from floating dead and half-burnt bodies of human beings and carcases of animals and wallowing of cattle on the one hand, and of municipal waste, often sheer faecal matter, both from the point and dispersed sources, on the other. The chemical pollutants are drained from industrial establishments in the cities located on the banks of the river system. While one can evolve a strategy to control pollution from point sources, the dispersed type of pollution is indeed very difficult to control as it involves very large numbers of people living on the banks. It is therefore also a socio-economic problem.

Though with tremendous resilience, this river system due to incessant pollutional load has lost its recuperative capacity in some stretches. About 480 km (out of over 2500 km) are dirty. Superimposed on this, are two more aspects. First is the drawal of water from the river which results in patches with very little flow during peak summer months so much so that one can easily wade or even walk through the river bed. Under the circumstances, even a small load of pollutants can be highly toxic and prove detrimental to the health of the river system itself and the people living on the banks. Then there is the problem of mass bathing (Kumbh Melas), a situation peculiar to India wherein the coliform counts jump several hundred per cent in a matter of days. Secondly, the catchment areas on account of denudation of forest cover, send down very heavy load of silt which is second only to the Yellow River in China.

A river system is not only a physical entity, it is a living dynamic and a continuing system; it is a resource. Obviously, any strategy that is applied to restore the system to its original condition must take this into account, and not merely resort to civil engineering work of using "Brick-mortar-drainpipe Technology", but take to a holistic approach. In this connection, I would commend a more or less forgotten paper by Prof. Rama (of TFIR) supported by Prof. Roger Revelle (California University). Rama has advocated a strategy which among other things amount to "Solution to Pollution by Dilution". It treats the whole river basin as a "sponge" and envisages drawal of water from aquifers and its storage in tanks, and then using the same purposefully in lean months. The empty aquifers would get re-charged during monsoon and the cycle would go on. This would also reduce the fury of floods.

Other connected aspects are dredging and use of the river system for navigation. Stabilisation of banks and use of "diara" lands are other dimensions.

"Mission on Ganga" is no less important than any of our prestigious projects on space and atomic energy. The dimensions of problems are unduly large as also complex. We need a visionary like Visvesvarya, Bhatnagar or Bhabha to comprehend and handle such a mission.

Tata Energy Research Institute, T.N. KHOSHOO
New Delhi.

Preface

DURING the fifth meeting of the Committee on Studies for Co-operation in Development in South Asia (CSCD) held in New Delhi in October, 1980, it was suggested that an exploratory overall investigation into the impact of human settlements and developmental activities on the Ganga river system in India, could be undertaken. A similar study might be carried out in Bangla Desh. A study group for the study in India was formed under the leadership of Prof. B.C. Haldar, Director of the Institute of Science, Bombay. This group contacted a large number of institutions and individuals, various state and central government departments, study groups and departments in universities, etc., concerned with a wide range of facets of work in the region of the Ganga river. Extensive correspondence was undertaken and discussions through personal visits were undertaken in many cases. The immense amount of data generated in the institutions and provided by them to this Study Group was screened meticulously, arranged methodically and evaluated for presentation in a comprehensive way. It was realized during the course of the work that, though the quantum of data on the river system was voluminous, there were numerous lacunae in the information that still needed to be plugged, e.g. the much talked-about phenomenon of deforestation in the Himalayas causing devastating floods in the river plains can be impressively brought about through mathematical models based on bare slopes, frequency of heavy rain incidents and runoff of soil and minerals. Similarly, the off-quoted process of exceptionally high self-purifying capacity of the Ganga river, will receive substantial evidence through studies on nutrient cycling processes in different parts of the river. This Report aims to identify such areas. It is hoped that the numerous organizations involved in different aspects of Ganga river studies, will be assisted by the present effort. Recently, the Planning Commission, Government of India, has initiated an extensive research and data collection programme, involving a large number of University Departments and National and Regional Laboratories, about the Ganga river ecosystem. This report is expected to form a starting point for their investigations.

An identical effort for part of the same river system by Bangla Desh, would lead to further appreciation of the environmental situation along the river system. It is hoped that the initiative of the Indian Council of World Affairs, New Delhi, who were responsible for motivating this Group to undertake the present effort, would be followed by a similar initiative in Bangla Desh. Combined appreciation of the output of these Groups would be a significant achievement in cooperation in the field of study of riverine environment, which would have enormous value for the people of the sub-continent.

March, 1983.

S.B. CHAPHEKAR
G.N. MHATRE

Acknowledgements

WE are grateful to the Indian Council of World Affairs, New Delhi, for giving us an opportunity to undertake this challenging task to bring together available information on the great Ganga river. Mr. Tarlok Singh and others from the Council have always been very co-operative in periodic discussions and in introducing us to many important sources of information. It speaks volumes for their understanding approach that they have been tolerant to our shortcomings.

We are thankful to the following persons and institutions for extending their co-operation by providing us with technical data available with them and for permitting us to use the same for incorporation in this Report. We are obliged to many of them for supplying us even their unpublished data. We are also grateful to Prof. R.L. Singh for permitting us to reproduce material and maps from his book, "India : Regional Geography".

Dr. Anil Agarwal, Director, Centre for Science & Environment, Nehru Place, New Delhi; Prof. R.S. Ambasht, Department of Botany, Banaras Hindu University, Varanasi; Shri Y.P. Bali, Project Officer, All India Soil & Land Use Survey, IARI Building, New Delhi; Dr. P.K. Banerjee, Environmental Officer, Department of Environment (Ecology), New Delhi; Shri S.G. Bhat, Library Documentation Cell, NEERI, Nagpur; Shri S.R. Chopra, Senior Research Officer, Education Division, Planning Commission, Yojana Bhawan, New Delhi; Dr. Nilay Choudhuri, Chairman, Central Board for the Prevention & Control of Water Pollution, New Delhi; Shri Das, Librarian, Central Water & Power Research Station, Khadkwasla, Pune; Shri Ghanekar, Director, Flood Control Division II, Central Water Commission, New Delhi; Shri M.G. Gogate, Asstt. Silviculturist, Forest Research Institute & Colleges, Dehra Dun; Dr. K.C. Jayaram, Deputy Director, Zoological Survey of India, Calcutta; Shri R.N. Kaul, Director, Forestry Research, Forest Research Institute & Colleges, Dehra Dun; Dr. A.K. Laal, Officer-In-Charge, Central Inland Fisheries Research Unit, ICAR, Bhagalpur; Dr. V.M. Meher-Homji, French Institute, Pondicherry, Shri K.S.S. Murthy, Advisor-Irrigation, Planning Commission, Yojana Bhavan, New Delhi; Shri P. Padmanabha, Registrar General & Census Commissioner for India, New Delhi; Shri S.N. Saraf, Educational Advisor, Education Division, Planning Commission, Yojana Bhavan, New Delhi; Mrs. Laxmi Vyas, Head of Geography Department, Elphinstone College, Bombay.

We would be failing in our duty if we do not make a grateful mention of the ex-Director of the Institute of Science, Prof. B.C. Haldar, and his staff for their unstinted administrative support to conduct the project and to the present Director of the Institute, who continued the co-operative support to the Project.

Drawings for the report are made by Shri G.D. Sawant while the help offered by all the colleagues in our Ecology Laboratory has been so varied that it is beyond our capacity to detail it. We also thank Indian Express, Newspaper, Bombay, and Prof. R.S. Ambasht of B.H.U., Varanasi, for sparing photographs.

We wish to put on record the active help given by late Prof. Haldar, Project Director, by way of advice and encouragement, to bring the report to completion, with continuous guidance through correspondence, in spite of his indifferent health, during the latter part of the work. It is unfortunate that he is not amongst us now, to witness the results of the effort.

March, 1983

S.B. CHAPHEKAR
G.N. MHATRE

Contents

List of Figures

List of Tables

List of Plates

Courtesy—Indian Express

BETWEEN PAGES 80-81

Photographs—Courtesy, Prof. R.S. Ambasht,
Department of Botany, Banaras
Hindu University, Varanasi.

CHAPTER I

Introduction

1.1. GENERAL INFORMATION ABOUT THE RIVER GANGA

GANGA is a perennial river formed by confluence of two smaller rivers at Devprayag. Bhagirathi is one of them originating at Gaumukh in Gangotri glacier, 3,129 m above sea level, while the other one is Alaknanda with her origin in Sapta Tal glacier. After gorging a distance of about 220 km in the Himalayas, it enters the plains at Hardwar and after meandering over a distance of about 2,290 km in the plains in the States of Uttar Pradesh, Bihar and West Bengal, it joins the Bay of Bengal through a large number of branches flowing in India and Bangla Desh (Fig. 1.1). The entire area comprises of wide geographical diversity. It consists of hilly terrain in the north up to Hardwar, low lying agricultural alluvial plains in the middle and deltaic region in the east before it meets the Bay of Bengal. Its drainage basin covers one of the most thickly populated regions of the world where the Indo-Aryan civilization flourished for many centuries. Ganga is formed by confluence of a number of tributaries like Yamuna, Kali, Karnali, Ramganga, Gandak and Kosi, all of which rise in the Himalayas and are mainly snow-fed. From the south, the tributaries joining the Ganga are the Chambal, Betwa, Tons, Ken, Sone, etc., which rise in the Vindhyan highlands of the Central part of India.

1.2. NEED FOR STUDY

The choice of the Ganga among the Indian rivers for environmental study has obvious reasons. It is the most vital life-line of the alluvial agricultural plains of North India and

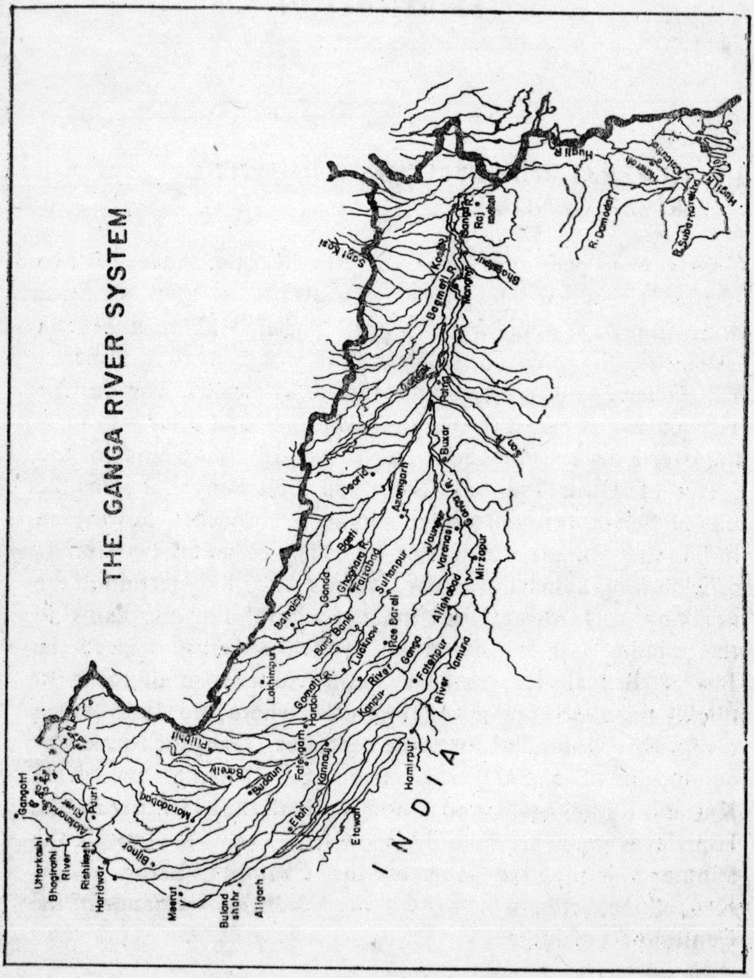
THE GANGA RIVER SYSTEM
INDIA
Gangotri
Uttarkashi
Bhagirathi River
Alaknanda River
Rishikesh
Haridwar
Pauri
Bijnor
Moradabad
Meerut
Buland shahr
Aligarh
Etah
Etawah
Budaun
Bareily
Pilibhit
Lakhimpur
Fatehgarh
Kannauj
Hardoi
Gomati R.
Kanpur
Lucknow
Bara Banki
Bahraich
Gonda
Ghaghara R.
Faizabad
Basti
Deoria
Sultanpur
Azamgarh
Rae Bareli
Ganga River
Fatehpur
Hamirpur
Yamuna River
Allahabad
Varanasi
Jaunpur
Mirzapur
Buxar
Son
Patna
Bagmati R.
Kosi
Sapt Kosi
Bhagalpur
Raj mahal
R. Damodar
Hugli R.
Howrah
Calcutta
R. Subarnarekha

FIG. 1.1

one of the most notable geographical dominants in the political history of the region. It has been an important element in religious, social, cultural and linguistic wealth of the area. It has been revered since the Vedic times in the Hindu scriptures.

1.3. THE BASIN

The approximate area of the total Ganga basin is 8,61,404 sq km (10,50,000 sq km including that in Bangla Desh) which is about 28% of the total Indian territory, lying along a total length of about 2,525 km in the northern part of the Indian subcontinent. The average annual runoff of the basin is estimated at about 4,93,400 million m^3 which is nearly 30% of the total water resources (16,70,000 million m^3) of the whole country. It has the largest gross sown area of nearly 58 million ha. It is a cradle of civilization for nearly 26% of India's total population (Ghosh, 1977).

Some important features such as irrigation potential, hydropower potential, construction of dams and reservoirs of the Ganga basin are given in Tables 1.1 to 1.3. These have contributed greatly to the development of the region.

TABLE 1.1 **Cultivable sown and irrigated areas in thousand ha of the Ganga Basin**

Cultivable area	Net sown area/ Gross sown area	Net irrigated area/ Gross irrigated area
60,161	46,047/57,673	16,049/19,496

From : Rao, 1975.

TABLE 1.2 **Hydropower potential of the Ganga Basin**

Power potential in MW	Installed capacity in MW	Effective capacity	Balance to be exploited
5,000	1,899	1,127	3,873

From : Rao, 1975.

TABLE 1.3 **Number of dams of 30 m height and above, built on the Ganga river in different States for the purposes of irrigation, hydel power, flood control, navigation, water supply and recreation**

States	Completed up to 1979	Under construction	Total
U.P.	7	7	14
Bihar	9	1	10
W. Bengal	1	nil	1

From : Register of Large Dams in India, 1979.

It is generally seen that development is accompanied by destruction. Many of the natural systems such as rivers have been affected due to increased human settlements and activities. Ganga is one such river which is facing grave dangers from increased human settlements and developmental activities all along its course.

The present report is a compilation of information on the physical and cultural settings along the Ganga river. The report is expected to provide a basis for assessment of the river quality, as affected by human activities. Since few investigations are reported so far in this field, the Report focusses attention on some such areas that require detailed investigations.

For the sake of convenience the river system has been divided into four zones, viz. :

(i) Zone above the Upper Ganga Plain, (ii) Upper Ganga Plain, (iii) Middle Ganga Plain, and (iv) Lower Ganga Plain.

Information on the following aspects has been compiled for each of the zones mentioned above :

(i) Geographical Location—(a) Administrative Divisions, (b) Physiography of the region.

(ii) The Physical Setting—(a) Geology, (b) Physiography,

(c) Climate, (d) Soil, (e) Drainage, (f) Natural Vegetation, (g) Ground Water, and (h) Physical Resource Base.

(iii) The Cultural Setting—(a) Population, (b) Urban Settlements, (c) Agricultural Economy, (d) Industrial Economy, and (e) Transport and Communications.

(iv) Problems in the region.

CHAPTER II

Zone Above Upper Ganga Plain

2.1. LOCATION

The area lies at the north tip of Uttar Pradesh between latitude 30° 0′N—31° 30′N and longitude 77° 45′E and 79° 30′E, covering about 10,336 km^2 area. The area bears the catchments of Bhagirathi, Bhillangana and Yamuna rivers. The north side is bounded by high ridge separating Uttar Pradesh from Himachal Pradesh and China. The southern boundary is formed by a ridge separating the Bhagirathi catchment starting from Devprayag in the east to Musoorie and Kasli in the west (Fig. 2 1). The region so defined incorporates within its bounds parts of the districts of Uttarkashi, Chamoli, Dehra Dun and Tehri Garhwal.

2.2. THE PHYSICAL SETTING

2.2.1. Geology

The area comprises of varied geolcgical formations belonging to different age groups. Most of the formations belong to Carboniferous, Permian, Archean, Tertiary and Crustaceous age. The major formations are Chail, Simla Slates, Garhwal series, Deoban Lime stone, Jaunsar, Krol, Chandpur, Nagnath, Granite and other crystalline formations.

2.2.2. Lithology

The rock types observed are also of varied nature. They are sandstone, quartzite, orthoquartzite, grit, conglomerate,

ZONE ABOVE UPPER GANGA PLAIN

ADMINISRATIVE DIVISIONS

Fig. 2·1

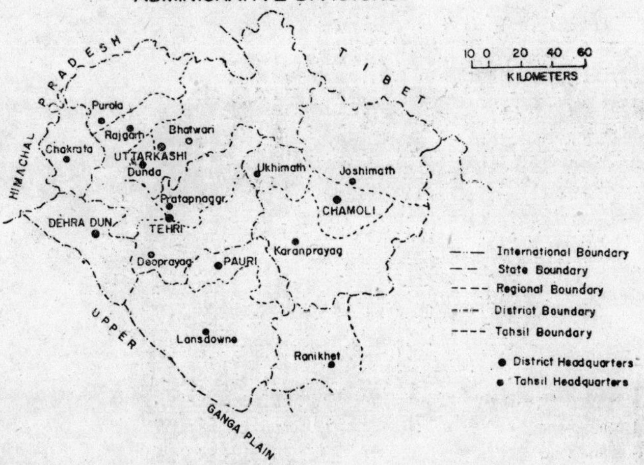

limestone, shales, slates, phyllite, gneiss, schist, granitic and other crystalline rock types.

2.2.3. Physiography

The entire area is hilly and undulating. Elevation varies between 400 m to 6,000 m in the region. The slopes also vary from steep to precipitous in nature (Fig. 2.2).

2.2.4. Drainage

The region is well drained by numerous rivers and rivulets. The drainage can be divided into two main systems : (1) The Ganga system, and (2) The Yamuna system.

Major part of the region is drained by the Ganga system covering the whole of Garhwal except western part of Uttar Kashi district and western part of Almora and Naini Tal districts. All the main rivers of Garhwal have a peculiar tendency to flow for some distance in structural troughs parallel to the mountains, but sooner or later they take an acute bend to flow in deep transverse gorges, at places hundred of metres in depth. The Bhagirathi and the Alaknanda originate from opposite sides of the Chaukhamba peaks (7,138 m). After

Fig. 2-2

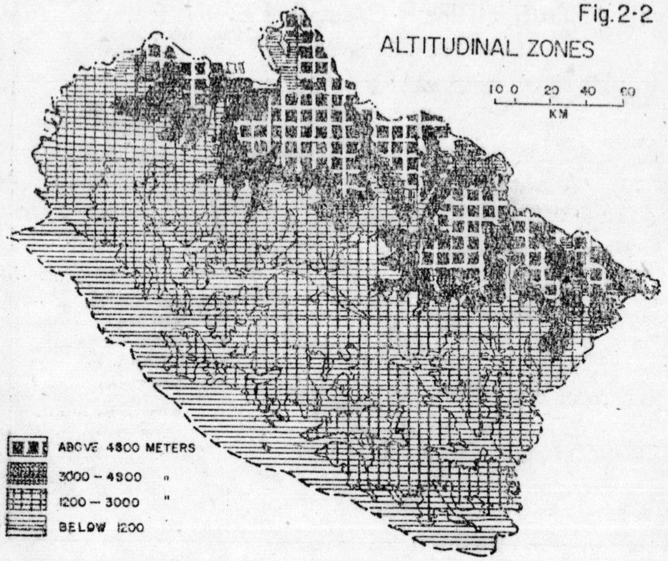

flowing in opposite directions longitudinally, they bend at Devprayag, thus taking a garland shape. The Bhagirathi and the Janhavi have cut awe-inspiring gorges through tourmaline granites constituting the central axis of the Himalayas.

Bhagirathi-Alaknanda Basin

Below Devprayag where the Bhagirathi and the Alaknanda meet, the river acquires the name of 'Ganga'. It emerges from the Himachal at Lachhman Jhula, flows through the *dun* and cuts across the Siwalik range at Hardwar. The Alaknanda is the main tributary of the Bhagirathi. Other tributaries of the Alaknanda are the Mandakini, the Pindar and the Dhawal Ganga which meet the main stream at Rudraprayag, Karnaprayag and Vishnuprayag respectively (Fig. 2.3).

The rivers have formed deep valleys, narrow or broad in accordance with lithology. In general, the dendritic pattern is common. Local radial pattern has developed around the hills and peaks. Trellis drainage is to be found in the areas of thrusts and local faults where main streams have their subsequent affluents at right angles.

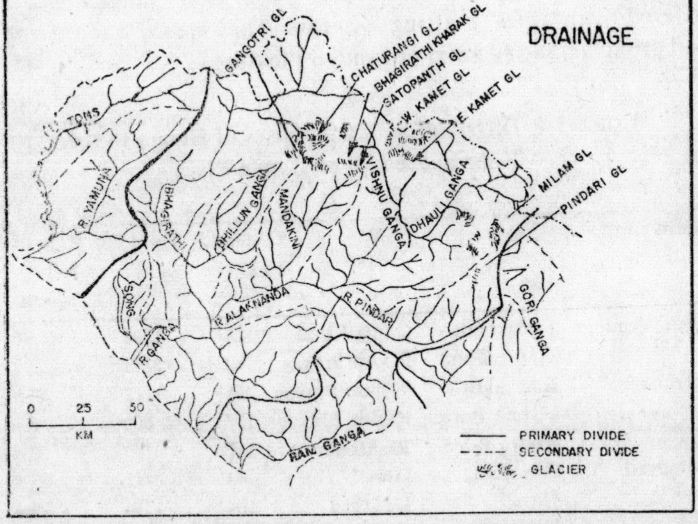

Fig. 2·3

All the main rivers make a steep descent in the first ten or twenty kilometres of their longitudinal profiles and afterwards their gradient is not so steep, as is clear from the Table 2.1.

2.2.5. Climate

Most of the region falls in the general category of temperate climate and the winters are very severe. The mean and annual temperatures recorded at different altitudes and in different climatic zones are given in Table 2.2. Annual rainfall ranges from 800-2,500 mm. 50% of the rain is in the form of snow in the higher reaches of the region. Areas located at 3,000 m and above are mostly snowbound throughout the year.

2.2.6. Soils

Four different soil types are found in this region. They are : red loam, brown forest, podzols and meadow. Red loam is found all along the slope of hills and along the bridges. The brown forest soils occur in most of the area which are also rich in organic matter. Soils are granular clayey or clayey micaceous or sometimes loamy in nature. Podzols occur on mild slopes in the pockets of the hills and shady places which are clayey to loamy in texture. They are dark brown to black in colour. Meadow soils occur where ground water level is high and are covered with thick matting of grass. They are dark in colour and granular, sandy loam or sandy in texture.

TABLE 2.1 **Distribution of tributaries of Ganga in the Zone above Upper Ganga Plain**

Name of the river	From	To	Distance in km	Total fall in m	Rate of fall in m/km
Mandakini	Kedarnath	Gauri Kund	14	1591	113.6
	Gauri Kund	Rudra-Prayag	70	1363	19.5
Alaknanda	Badrinath	Vishnu Prayag	38	1818	47.8
	Vishnu Prayag	Rudra Prayag	117	600	5.1
	Rudra Prayag	Rishikesh	141	270	1.9
Bhagirathi	Gaumukh	Harsil	43	1445	33.6
	Harsil	Dharasu	104	1394	13.4
	Dharasu	Rishikesh	122	690	5.6

From : Singh, 1971.

TABLE 2.2 **Annual mean temperature at different altitudes and in different climatic zones in the Zone above Upper Ganga Plain**

Climatic zones	Altitude in metre	Temperature °C Annual mean
Tropical Zone	300— 900	21.1
Warm Temperate Zone	900—1800	18.9
Cold Temperate Zone	1800—2400	13.9
Cold Zone	2400—3000	10.3
Alpine Zone	3000—4000	4.5
Glacial Zone	4000—4800	Ten months below zero, two months between 2.2 & 3.9
Perpetually Frozen Zone	Above 4800	Cold desert—no vegetation

From : Singh, 1971.

2.2.7. Natural vegetation

The entire area is very rich in forests (Fig. 2.4). The important forest types (as per Champian and Seth's classification) which occur in the area are :

(1) Northern tropical dry mixed deciduous forests,
(2) Himalayan subtropical pine forests,
(3) Lower Western Himalayan temperate forests,
(4) Upper Western Himalayan temperate forests, and
(5) Montane Himalayan moist temperate forests.

Fig. 2·4

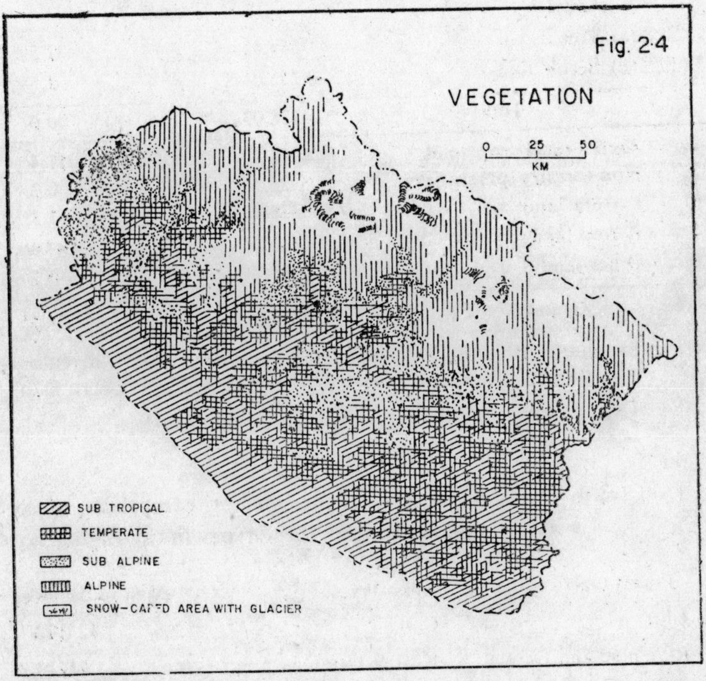

2.2.8. Physiognomy

Land use patterns and different forest types observed in the region are given in Tables 2.3 and 2.4.

2.3. THE CULTURAL SETTING

2.3.1. Population

In a mountainous region like the Himalayas, topography and

climate exert more influence on the distribution of population than do other factors. Maximum concentration of population is found in the middle zone along the lower parts of the Bhagirathi, the Alaknanda, the Yamuna, the Ramganga and the Kosi valleys.

TABLE 2.3 **Vegetation types near the origins of Bhagirathi, Bhillangana and Yamuna rivers**

No.	Land use class	Area sq km	% of total area
1.	*Forest land*		
	(a) Tree forest	3873.02	37.5
	(b) Scrub forest	255.01	2.5
	Total	4128.03	40.0
2.	Agricultural crop land	1183.42	11.4
3.	Non-forestry plantation	23.91	0.2
4.	Pasture land	338.69	3.3
5.	Barren land	4586.26	44.4
6.	Other land	75.71	0.7
	Grand Total	10336.02	100%

From : Bhagirathi, Bhillangana & Yamuna Catchments, Vol. I, Inventory Results, Pre-Investment Survey of Forest Resources, Dehra Dun, Deptt. of Agriculture, U.P., 1976.

TABLE 2.4 **Types of forests, plant species comprising these types, and area under these types in the region**

Forest type	Species	Area in hectares
1. Chir	Chir	122172
2. Blue pine and Deodar	Kail & Deodar	21210
3. Fir & Spruce	Fir & Spruce	21635
4. Oaks	Ban, Mohru & Kharsu Oak	169259
5. Low land hard wood	Low land hard wood	9757
6. Upland hard wood and mixed forests	Upland hard wood and mixed coniferous	43269

From : Bhagirathi, Bhillangana & Yamuna Catchments, Vol. I, Inventory Results, Pre-Investment Survey of Forest Resources, Dehra Dun, 1976.

The density per km^2 ranges from 24 in the hilly region of Uttarkashi district to 245 in Dehra Dun district, (Census, 1981). The Siwaliks, a major part of duns and the foothills of the Himachal were covered with dense forests where the density of population ranged between 5 and 10 persons per km^2 (1961 Census). In the forest clearings however, high densities of population are found even in this zone.

2.3.2. Trend of growth

During the past, the region has experienced relatively low population growth due to the preponderance of natural calamities, social customs such as polyandry and migration to other areas of young people in search of employment. Attacks of wild animals, avalanches, floods, landslides, diseases, etc., also claimed a fairly high number of lives. The region has recorded about 80% increase in its total population during 1901-1961 and except during 1911-1921 the growth of population has been more or less steady in those six decades (1901-1961). The growth rate for 1971-1981 is around 27% for the region.

About 17% of the population of the region is urban. The region shows wide subregional disparity in the distribution of the urban population. Whereas Tehri Garhwal shows 4.17%, Uttarkashi shows 7.0%, Chamoli shows 8.0% and Dehra Dun has 49% urban population which is the highest for the region (Table 2.5).

Annual growth rate for urban population for the same decade for different districts ranged from 3.7% in Dehra Dun to 13.6% in Chamoli.

The working population in different districts through which the Ganga and its tributaries flow, range from 32% in Dehra Dun district to 54% in Uttarkashi district. The other two districts, viz., Chomoli and Tehri Garhwal showed 44.59% and 41.59% working populations respectively in the 1981 Census of India.

The distribution of main workers in different sectors such as cultivators, agricultural labourers, household industry and other workers showed that except in Dehra Dun district which showed almost 60% other workers and 3% workers in other

TABLE 2.5 **Urban distribution of districtwise population in the Zone above Upper Ganga Plain**

District	Density of population km^{-2}	Population, 1981		Percentage of Urban to total population		Percentage growth rate, 1971—1981	
		Total	Urban	1971	1981	Total	Urban
Uttarkashi	24	190,571	13,263	4.07	6.96	+28.93	+120.32
Chamoli	40	364,287	28,780	4.17	7.90	+24.51	+135.79
Tehri Garhwal	112	493,245	20,580	2.65	4.17	+24.12	+ 95.57
Dehra Dun	245	757,259	372,652	47.08	49.21	+31.17	+ 37.12

From : Appendix I in Paper 2 of 1981 Rural-Urban Distribution and Appendix III in Paper 3 Workers and Non-workers, Census of India, Series 1, 1981.

TABLE 2.6 Distribution of main workers—cultivators, agricultural labourers, household industry, and other workers in different districts situated in the Zone above Upper Ganga Plain

District	Percentage of Main workers in Agriculture, Household Industry & other workers to Total Workers							
	Cultivators		Agricultural		Household Industry		Other workers	
	1971	1981	1971	1981	1971	1981	1971	1981
Uttarkashi	85.90	79.67	1.00	0.57	0.73	1.33	12.37	18.43
Chamoli	86.52	80.76	0.44	0.38	1.71	2.10	11.33	16.76
Tehri Garhwal	90.96	84.41	0.55	0.50	1.09	0.91	7.40	14.18
Dehra Dun	28.38	29.02	8.19	7.99	2.47	3.10	60.96	59.89

From : Appendix II in Paper 3 of 1981, Workers and Non-workers, Census of India, Series 1, 1981.

industry, the other three districts showed very low percentages of these types of workers reflecting that the agriculture is by far the most important occupation in the region (Table 2.6).

2.3.3. Urban settlements

The zone has eleven towns and one urban agglomeration, i.e., Hardwar (Table 2.7). Towns occcupy four types of sites in this region.

TABLE 2.7 **Population of urban agglomerations, cities and towns situated in the Zone above Upper Ganga Plain**

Urban agglomeration/ cities/towns	Population 1981	Growth	
		1961-71	1971-81
* 1. Chamoli Gopeshwar (N.A.)	9,734	—	+ 53.19
**2. Joshimath (N.A.)	8,178	—	+ 39.75
**3. Rudraprayag (T.A.)	1,331	—	—
**4. Karnaprayag (N.A.)	3,767	—	—
5. Devprayag (N.A.)	1,727	+ 4.88	+ 13.10
6. Uttarkashi (M.B.)	10,043	+124.88	+ 66.83
7. Tehri (M.B.)	12,248	+ 21.56	+123.50
8. Narendra Nagar (M.B.)	3,591	+ 46.45	+ 50.25
9. Pauri (M.B.)	13,610	+ 18.63	+ 53.30
10. Srinagar (M.B.)	9,205	+ 83.64	+ 65.38
11. Rishikesh (M.B.)	29,100	+ 61.52	+ 64.91
12. Hardwar (U.A.)	146,186	+ 32.22	+ 59.99

N.A.=Notified Area Committee, T.A.=Town Area

M.B.=Municipal Board, U.A.=Urban Agglomeration.

* Towns treated as such for the first time in 1971 and continuing as such in 1981.

**Towns treated as such for the first time in 1981.

From : Appendix II in Paper 2 of 1981 Rural-Urban Distribution, Census of India, Series 1, 1981.

1. *Ridges*—Pauri
2. *River terraces*—Joshimath, Srinagar and Uttarkashi.
3. *Confluences*—Tehri, Devprayag, Rudraprayag, Karnaprayag.
4. *Gate towns*—Rishikesh, Hardwar.

Religious and administrative functions play a vital role in the process of urbanization in this region. According to Hindu philosophy a spot of beauty is no place for social enjoyment or self-indulgence. It is the place for self-restraint, for solitary meditation which leads the mind from Nature up to God. In India, nowhere this concept is exemplified more than in the Himalayas. Thus, religious cities have developed around the famous holy shrines which have been set up by saints at sites of exquisite natural beauty. Here, the devotees could perform their penance and meditation in a calm, serene and sublime atmosphere. Devprayag, the confluence town, is highly congested in a narrow triangle with its apex at a lower level and base at a higher level. Most of the towns of the old Tehri State have been founded by their rulers and are identified after their names, e.g., Narendranagar, Joshimath, etc. The latter has now become an important military centre. Most of the towns depict a linear pattern as they have developed on ridges or terraces.

Between Rishikesh and Hardwar there is a fragmented ribbon of urban settlements. Parts of this ribbon are more developed with the establishment of an antibiotic factory at Virbhadra, Bharat Heavy Electricals Ltd. at Ranipur and a military centre at Raiwala. The whole strip (about 25 km long) along the Ganga is strung with religious and educational institutions.

Hardwar which is the only urban agglomeration situated along the Ganga river in the region showed an increase of 60% in its growth in the decennial period ending 1981. During 1961-71, the increase was 32% (Table 2.7). 27% (i.e., 39,591) of the population constitutes the work force. The rest of the population (106,565) is of non-workers, (Table 2.8). The distribution of main workers in different categories showed that the number of 'other workers' was the highest as compared to all the other categories, viz., household industry, agricultural

TABLE 2.8 **Distribution of population by workers and non-workers in Hardwar, an urban agglomeration**

Total main workers	Workers as percent to total population	Cultivators	Agricultural labourers	Household industry	Other workers	Marginal workers	Non-workers
39,591	27.08	392	1,781	1,148	37,270	30	1,06,565

From : Appendix IV in paper 3 of 1981—Provisional Population of Total Workers and non-workers, Census of India, 1981.

labourers and cultivators. These figures indicate the extent of total urbanism of the place with very low agricultural activity.

2.3.4. Agricultural economy

About 40% of the area is under forest. Only 11.4% of the total area is under agriculture (Table 2.3). The cropping pattern in the region reflects the varying conditions of altitude, climate and soil on one hand and different agronomic and cultural practices on the other. These traditional practices such as elaborate terracing, crop rotations and gul irrigation system, the last one also providing water for small water-mills, have evolved over centuries. The physical environment has infused a high degree of adaptive skill and uncommon physical vigour among the people to cope with the inhospitable environment.

A typical valley section of the region exhibits three distinct categories of land use (Pant, 1935) :

(i) In the 'Katil' or 'Khil' land, hoe cultivation is practiced with a rather standard intermittent rotation of three crops in five years. After a few years of intermittent cultivation, such land is finally terraced and regularly cultivated. 'Madua' (Mandua) or 'Kodo' (*Elucine corucana*) is the most important crop favoured in this form of cultivation. Other crops are 'Jhangora' (*Oplismenus framentaceous*) and 'Marsa' or 'Chua' (*Amaranthus frumentaceous*).

(ii) Permanently terraced but unirrigated cultivation is found in the 'Upraon' or 'Ukhar' lands with 'Madua', 'Jhangora' and 'Chua' as main crops.

(iii) 'Talaon' land is mostly devoted to paddy in lowlying irrigated areas, called 'sera' which are invariably double cropped, while the other is of 'Tanr' type used for broadcast paddy, sometimes irrigated by 'gul' system.

From ecological point of view the distribution of crops follows some sort of vertical zonation (Kaushic, 1962). Wheat is widely distributed and even in the cold zone at altitude of 2,400-3,600 m, it is grown as a summer crop. Rice is confined only to lower and middle zones between altitude of 300 and 1,800 m. Millets are highly diffused except in the cold zone. While barley is grown in the upper zone (1,800-3,600 m), sugarcane thrives only in the lower Dun (300-600 m). Potato thrives everywhere except in the lower altitudes.

2.3.4.1. Rotation of crops

'Three crops in two years' rotation is the most important practice prevalent on 'upraon' lands. For the sake of these crop rotations, the cultivated area of a village is divided into several parcels (Sar). Each 'Sar' is allotted a definite crop rotation. The 'Sar' growing rice is called 'Dhan-ki-Sar', another growing 'madua' is known as 'madua-ki-sar'.

2.3.4.2. Crop combinations

Millets is the first ranking crop in the region while the second position is taken by wheat leaving rice in the third position. Among all the crop combinations, the millet-wheat-rice (MWR) combination occupies the largest area. The distributional pattern of each of these crops reflects differences in the availability of water, nature of terrain and method of cultivation. Dominance of rice is found in the 'talaon' lands, while that of millets in the drier 'upraon' and 'katil' lands.

Horticulture occupies a significant place in certain favourable areas of the region. Climate and soils in the altitudinal zones of 1,200-2,400 m are suitable for the cultivation of apples, pears, citrus fruits, cherries, plums, walnuts and apricots In Uttarkashi, there are 1,716 hectares under fruit cultivation. People are taking keen interest in the development of horticulture.

2.3.5. Industrial economy

In addition to rich hydropower potential, the region derives its industrial resources from forests, livestock, agriculture including horticulture and minerals, to permit industrial development. The region has developed a variety of small-scale and cottage industries based on cheap and abundant hydroelectric power and manpower of the region.

The region has great attractions for pilgrims, tourists and mountaineers. It has prominent shrines and tourist spots of everchanging resplendent scenic beauty with a wide range of flora and fauna. More than 2 lakh people visit the holy shrines every year and spend between 10 and 20 million rupees.

The climate, topography, vegetal cover, perennial flow of streams, absence of other competitive fuels such as coal and oil are all favourable factors for the development of water power in this region. According to Kaushic (1965) the Alaknanda and Bhagirathi basins possess 6,00,000 kw and 7,00,000 kw power potential respectively.

The mining activity is underdeveloped and the detailed exploratory work for minerals is in progress. The absence of coal, inaccessibility and non-development of water power have retarded the development of mining in this region.

Sugar mills are concentrated in various towns of the 'Dun' of Dehra. Miniature bulb factories, textiles mills and lime kilns are some of the main industries. Among the small industries are tea-processing, paper and pulp and manufacturing of some scientific instruments. An antibiotic factory at Virbhadra (Rishikesh) and Bharat Heavy Electricals Ltd. at Ranipur (Hardwar) are the only big factories worth the name in the region.

It has already been mentioned that the region is rich in forests. The largest proportion of the area under forests is in the districts of Uttarkashi (83.9%) followed by Tehri (68.6%) and Dehra Dun (50%). Thus, development of forest-based industries is inevitable. Most of the cottage industries are directly related to the products of ringal bamboo, forest products and livestock. The important household industries are wool weaving, wicker work, rope-making, leather working, carpentry, distilling country liquors, etc.

2.3.6. Transport and communications

Though strategic importance of the northern frontiers has accelerated the development of transportation and communication in this region since 1962, the isolation of some parts is still unbroken. Horticulture and commerce have been stimulated wherever modern roads have reached. The existing transport arteries also perform an enormous task of handling a huge rush of pilgrims and tourists during the summer and post-monsoon periods.

From the Rishikesh railhead start two famous mountain roads, one to Badrinath and the other to Gangotri. However, the roads are liable to be blocked by landslips or landslides during the rainy season. No air services are available in the region excepting for military purposes (Fig. 2.5).

2.4. PROBLEMS IN THE REGION

The region is endowed with vast resources of forests, scenic beauty, agricultural and horticultural products and livestock. Such a natural wealth is affected by the recent increase in human populations and their activities.

Due to hilly terrain and undulating topography the distribution of industries is uneven. At the same time, their scattering has resulted in no obvious instances of pollution—an offshoot of industries.

However, the region faces the peril of urbanisation due to increase in the human population in the last decennial period. The region has shown an increase of 27% in the total population, which is causing stress on the natural resources such as forests which are getting depleted at a faster rate. The net growing stock of different forests in the catchments of Bhagirathi, Bhillangana and Yamuna rivers is given in Table 2.9. Unless there is a check on the consumption of this resource it will get destroyed in a very short period. Summary wood consumption in the same area in the year 1972-73 shows total wood consumption of 1,61,340 m^3 (Table 2.10). The latest figure is expected to be much higher bearing in mind the growth in population and industries and consequent clearings of forests. Though these figures are not readily available, the awakening of the local population to the problem of indiscriminate deforestation as exemplified by a very powerful 'chipko' movement, indicates the severeness of the situation.

An increase in the rate of deforestation has created a great impact on the stability of the region. The Ganga has flooded every year during the last decade affecting the regions situated

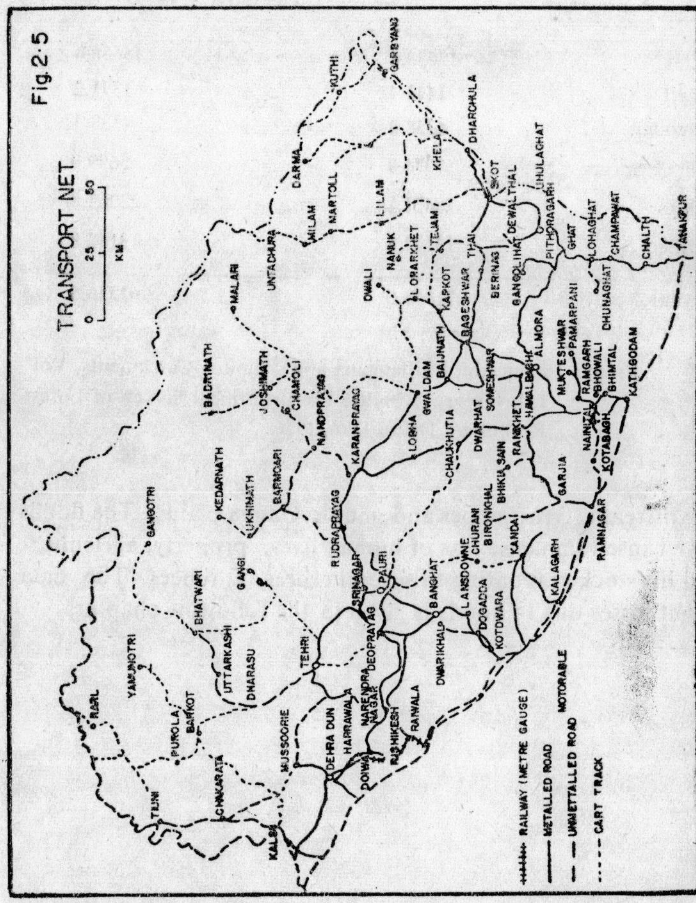
Fig. 2.5
TRANSPORT NET
0 25 50
KM
RAILWAY (METRE GAUGE)
METALLED ROAD
UNMETTALLED ROAD
MOTORABLE
CART TRACK
KALSI
TIUNI
CHAKARATA
PUROLA
BARKOT
RARL
YAMUNOTRI
MUSSOORIE
DEHRA DUN
DOIWALA
HARRAWALA
RISHIKESH
RAIWALA
NARENDRA NAGAR
DHARASU
UTTARKASHI
BHATWARI
GANGOTRI
GANGI
TEHRI
DEOPRAYAG
SRINAGAR
PAURI
DWARIKHAL
KOTDWARA
DOGADDA
LANSDOWNE
BANGHAT
KALAGARH
CHURANI
KANDAI
BIRONKHAL
BHIKIA SAIN
RAMNAGAR
GARJIA
KOTABAGH
NAINITAL
KATHGODAM
BHIMTAL
BHOWALI
RAMGARH
MUKTESHWAR
PAMARPANI
CHUNAGHAT
DHUNAGHAT
LOHAGHAT
CHAMPAWAT
CHALTH
TANAKPUR
GHAT
PITHORAGARH
ALMORA
HAWALBAGH
RANIKHET
DWARHAT
CHAUKHUTIA
SOMESHWAR
GANGOLIHAT
BERINAG
THAL
DEWALTHAL
SKOT
DHARCHULA
KHELA
UNULACHAT
BAGESHWAR
BAIJNATH
GWALDAM
KAPKOT
TEJAM
LOHARKHET
NANUK
DWALI
LORHA
KARANPRAYAG
RUDRAPRAYAG
UKHIMATH
KEDARNATH
BARMOARI
NANDPRAYAG
CHAMOLI
JOSHIMATH
BADRINATH
MALARI
UNTACHURA
MILAM
MARTOLI
LILAM
DARMA
KUTHI
GARBYANG

TABLE 2.9 **Net growing stock of wood in the catchments of Bhagirathi, Bhillangana and Yamuna rivers**

Species	Volume in '000 m^3	
	Gross	Net (after deduction of cull)
Chir	15658.4	15658.4
Kail	1443.4	1371.2
Deodar	4338.4	4338.4
Fir & Spruce	7074.9	5659.9
Oaks	28421.1	22065.8
Others	13142.8	13142.8
Total	71079.0	62236.5

From : Bhagirathi, Bhillangana and Yamuna Catchments, Vol. I, Inventory Results, Pre-Investment Survey of Forest Resources, Dehra Dun, 1976.

downstream, viz., upper and middle Ganga plains. The floods have caused immense loss of human lives, property, agriculture and livestock amounting to several crores of rupees. The data about losses due to floods is given in the following chapters.

TABLE 2.10 **Summary wood consumption in the catchments of Bhagirathi, Bhillangana and Yamuna rivers in the year 1972-73**

Item	Species (Volume in m³)					
	Chir	Kail	Deodar	Fir/Spruce	B.L.	Total
House construction & repairs	17279	2765	10367	—	4147	34558
Agricultural implements	2565	—	855	—	13680	17100
Fuel (Recorded)	625	78	78	260	14309	15350
Other domestic uses	3223	—	—	—	3223	6446
Industries						
(a) Furniture	70	70	325	—	118	583
(b) Packing case	26520	—	—	6630	—	33150
(c) Hydel projects	—	—	8908	—	7809	16717
(d) Other departments	520	2	15	—	23	560
Total internal consumption	50802	2915	20548	6890	43309	124464
(i) Railways	22706	42	766	3285	—	26799
(ii) D.G.S. & D.	2446	57	1048	6525	—	10076
Total of committed export	25152	99	1814	9810	—	36875
Grand Total	75954	3014	22362	16700	43309	161339

From : Bhagirathi, Bhillangana and Yamuna Catchments, Vol. I, Inventory Results, Pre-Investment Survey of Forest Resources, Dehra Dun, 1976.

CHAPTER III

Upper Ganga Plain

3.1. LOCATION

THE Upper Ganga plain is a part of the Great Plains lying approximately between the Yamuna in the west and 100 m contour in the east (73° 3′ E—82° 21′ E and 25° 15′ N—30° 17′ N) covering about 1,49,029 km^2 (about 57%) of the area of the State of Uttar Pradesh. The region is delimited in the north by 300 m contour which separates it from the Garh—Kum Himalaya west of the Sarda while the international boundary of Nepal marks the limit towards the east. In the south the Yamuna demarcates its border with the Bundelkhand. The region markedly differs in physical, social and economic characteristics from the northern and southern bordering regions but it is devoid of any such limits on the west and the east where the lie of the land with imperceptible physical, climatic and economic variations seldom provide any landmark to put a precise divide between its counterparts the Punjab Plain on the west and the Middle Ganga Plain on the east.

The region (550 km E-W and 380 km N-S) incorporates the Divisions of Meerut (excluding Dehra Dun district), Agra, Rohilkhand, Lucknow, Allahabad (excluding the tahsils of Handia, Phulpur, Meja and Karchhana) and Faizabad (excluding the tahsils of Utraula, Balrampur, Faizabad, Tanda, Akbarpur, Sultanpur, Kadipur and Patti) and southern part of Haldwani tahsil of Naini Tal, Kumaun Division (Fig. 3.1). Culturally, the region presents an amalgam of prehistoric to present culture, while politically, it has enjoyed relatively greater stability, though with occasional interruptions during transitions induced by changing dynasties. It is then that the

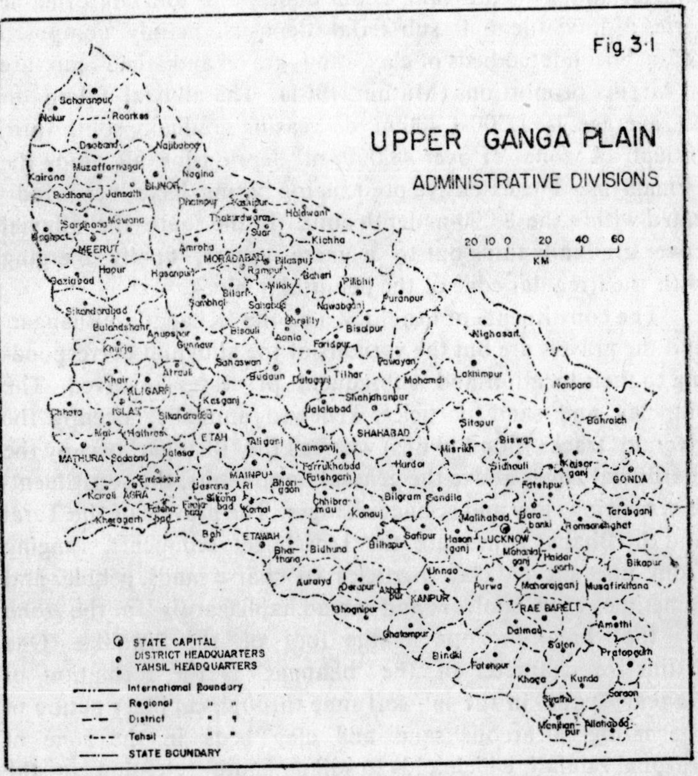

local aspirants like the 'subedars' of Awadh, tried to assert their supremacy.

3.2. THE PHYSICAL SETTINGS

3.2.1. Geology

An almost imperceptible change in elevation and uniform surface materials are the two noteworthy features in the physiognomy of the region. It forms part of the alluvial asymmetrical Indo-Ganga trough with a definite divide formed by the subterranean Delhi Ridge, a protrusion of the Peninsular Block. The Pre-Tertiary river-borne debris from the Peninsula, later supplemented rather more vigorously by the Upper and Post-Tertiary Himalayan debris that is yet to undergo intense compaction, constitute by and large, the alluvial fillings. Alluvium is one continuous and conformable series of fluviatile and sub-aerial deposits, mainly composed of unconsolidated beds of clay, sand, gravel and their mixture in varying proportions (Mathur, 1961). The alluvial filling on an average is 1,300-1,400 m decreasing gradually southward, though a zone of over 8,000 m depth runs all along the Himalayas. The extensive pre-existing basins have been indicated within the 8,000 m depth zone. In the south the alluvial cover gradually thins out to a mere veneer, finally merging with the irregular edge of the peninsular block.

The constituents of lithology, i.e., the 'Khadar', 'Bhangar' and the gravels are but the variants of the alluvium corresponding to their location and subsequent processes involved. The silt, clay and sand particles, renewed annually, occupy the riverain tracks (flood plains) whereas the 'Bhangar' occupy the interfluvial zones above the general flood limits, the constituents experiencing slow and secular changes. The gravels of the Tarai and the Bhabar zones are the loosely set sediments, ranging from fine silt or clay particles to coarse sands, pebbles and sometimes even boulders and spread haphazardly in the zone of the break-in-slope at the foot of the Siwaliks. One distinctive character of the 'bhangar' is the formation of 'kankar' pans in the sub-soil zone through capillary action of alternating calcareous sand and clay beds in the zone of seasonal rainfall, which adds to soil moisture retention in the subsoil zone.

3.2.2. Physiography

This seemingly featureless plain lacks topographic prominences and the monotony of the physical landscape is broken by the Tarai Bhabar submontane belt and on micro level by the river bluffs, levees, dead arms of the river channels, the ravines and the river channels themselves (Fig. 3.2). Paradoxically the axis of the topographic trough lies nearer the peninsular block or along the Ganga which traverses the area in a south-southeasterly direction with an average gradient of about 24 cm km^{-1}. The gradient, however, is higher (about 100 cm km^{-1}) in the upper 140 km of its course but much lower (about 10 cm) in the remaining part. Thus, there is an imperceptible tract adjacent to the foothills where the slope is higher and has resulted in the preponderance of numerous small streams. The latter has assigned a somewhat medium to fine texture to this part. The southern counterparts particularly north of the Ganga, are characterised by the sluggishly flowing streams like the Ramganga and the Ghaghra, studded with ox-bows, sandy stretches, etc. Even the relative relief is less pronounced here.

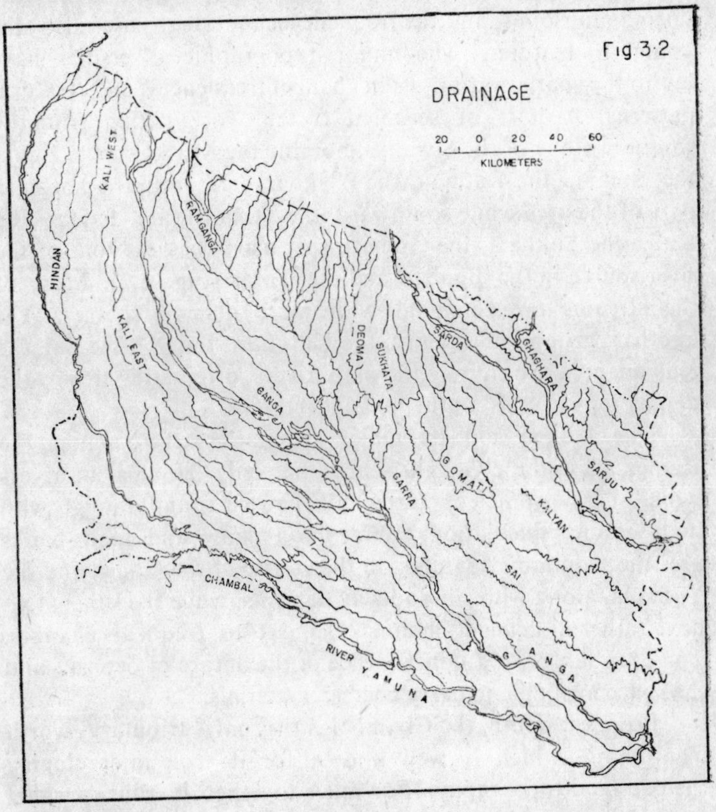

3.2.3. Drainage

The region in general is a part of a well integrated drainage system of the Ganga, although the two important tributaries, the Ghaghra and the Gomati, join the master stream in the Middle Ganga Plain. Almost all the streams follow a NW-SE course concomitant with the lie of the land. The extremely gentle gradient almost all over the region restricts the degradational activities of the streams resulting in the near parallel courses and the acute angle junctions of the tributaries with their master streams. Confluence of Ramganga with Ganga is a typical example. This feature imparts the region a pinnate drainage, an extreme case of the dendritic pattern on macro level. Braidings, especially of the Yamuna, the Ganga, the Ramganga and the Ghaghra, the sandy shoals often liable to innundation during monsoons, and the frequent meanderings, are also the common features. The minor topographic diversities have led to the spatial variations in channel frequencies and textural patterns. A host of seasonal torrents originating from the southern slope of the Siwaliks, notably the Muskara, the Punja the Solani, the Ratmau, the Pilakhar, etc., drain the northern part of the area while southward the streams are fewer. The Kali, the Sukhata, the Gomati, etc., mark the next zone of the river source in the 'Bhabar' and the 'Tarai' (Fig. 3.2). Most of the streams are perennial with more defined courses. Yet another group of streams, the Kali East, the Sot, the Sai, the Kalyan, etc., are entirely the plain rivers originating from the depressions of 'tals' in the 'bhangar' tracks.

The Ganga and its major tributaries, the Yamuna, the Ramganga and the Ghaghra are the only Himalayan rivers which carry sufficient water all the year round, though with high seasonal fluctuations. Wide flood plains and high banks are the common features in the course of the Ganga and the Yamuna, along with silt and clay deposits, while the other two have rather ill-defined channels subject to frequent changes. This probably is a consequence of the nature of deposits and the space available to carry coarser materials.

From the south, the Chambal is the only tributary worth mentioning, which is well known for its role in developing the bad land topography. For miles together it runs parallel to the Yamuna, before joining it.

Regionally, most of the Awadh and the Sarayupar plain is drained by the Ghaghra and the Gomati while the Ramganga collects most of the drainage of the Rohilkhand Plain. The Doab has few streams worth the name joining the Ganga and Yamuna.

3.2.4. Climate

It is a sub-humid region between the dry Punjab Plain and the humid Middle Ganga Plain, within the vast monsoonal regime of the Great Plains. It naturally partakes the characteristics of the two adjoining regions. With comparatively greater incidence of winter rain, the region distinguishes itself from the Middle Ganga Plain. Micro level diversities are no doubt introduced by the proximity of the Himalayas.

The average weather conditions emerging out of the combined effect of the various elements lead to the recognition of four well marked seasons, i.e., the Hot Summer, the Wet Summer, the Pre-Winter transition and the Winter. The pre-monsoon (March—May) showers are meagre (below 5 cm). The low relative humidity and meagre rainfall accentuate the impact of 'loo' which at times is associated with heat waves when temperature shoots up to exceptionally high levels. The rainy months account for over 90% of the total annual rainfall. The monsoon rainfall as also the annual rainfall decreases westward as well as southward (Gonda 113 cm, Bahraich about 100 cm, Allahabad 90 cm, Saharanpur 75 cm, and Agra 68 cm). The winter conditions settle in from November, marked by fall in temperature and prevalence of dry and chilly westerlies. These are occasioned by the western depressions associated with rain and cold waves and registering temperatures below freezing at some places. The average temperature is lowest in January (below 20°C) which spatially decreases west and northward. The incidence of winter rains in the region is relatively more than the Middle Ganga Plain (3.4%). Even this meagre rainfall, associated with the dampness of atmosphere (relative humidity 55% to 75%) and lower potential evapotranspiration due to the low temperature, proves effectively beneficial to the 'rabi' crops.

The average annual rainfall varies between 50 cm and 140 cm with an uneven spatial distribution. The Doab, west of Kanpur except the north eastern part, experiences less than 80 cm rain. The characteristic features are the westward and southward decline in the total rainfall and the incidence of double maxima (January-February and July-August) though not comparable in magnitude.

3.2.5. Soils

In this region of almost uniform topography and lithology, the soils are by and large, homogeneous. The alluvial soils at different parts are observed to have variants like 'Usar' and 'Bhur', depending on the drainage conditions, mechanical and chemical constitutions, climatic characteristics, etc. There are also minute variations in certain properties, according to which there are types like the 'Khadar' and the 'Bhangar' that are quite widespread.

The 'Khadar' soils are relatively rich in plant nutrients and occupy the narrow frequent siltation tracts in the flood plain of the rivers. Neutral to alkaline in reaction (pH 6 8), these are deficient in organic materials specially organic phosphorus and are sandy to loamy in texture. The Ganga 'Khadar' soils have immature profiles with sandy to silty loam texture, lack of concretion, fair proportion of lime and other soluble salts and are alkaline in reaction (pH 8). They also show imperfect drainage (Puri, 1960).

The 'bhangar' soils are more extensive in areal spread, occupying the interfluvial zones. Relative consistency in the factors responsible for the soil development has provided it a somewhat mature profile with varying colours. In general the soluble salts and lime are low and the soils show neutral to slightly acidic reaction except in waterlogged soils with pH 6 to 7.5. In the proximity of the Ganga these are loamy to sandy loam in texture.

Among the variants worth noting are the 'usar' ('reh') soil in a sizable tract in the Ganga-Ghaghra Doab from Sultanpur, Pratapgarh through Rae Bareilly, Lucknow, Hardoi, Shahjahanpur, etc., contiguous with the tract of the Middle Ganga Plain. The canal irrigation aggravates the 'Kallar' problem. Predominance of sodium provides the 'Usar' soil a neutral to acidic character whereas in Lucknow tract additional materials like calcium, makes it highly alkaline.

Another variant, the 'bhur' or the sandy river deposit, is highly localised in Ramganga tract and in the narrow belt along the Ganga. The soil is more sandy in texture and workable economically only with irrigation.

Slightly different from the 'khadar' and the 'bhangar' is

the submontane soil where two physiographic units, the 'bhabar' and the 'tarai', are bedded with texturally different soils. The 'bhabar' soil is sandy to gravelly, highly porous and aerated, and has lower moisture retaining capacity, while the 'tarai' zone is provided with rich clayey soil, with some proportion of fine sand, moisture and rich humus. The two zones are thickly forested. The deforestation in the tarai brings about considerable change in the soil texture, humus content, etc.

3.2.6. Natural vegetation

Originally the Upper Ganga Plain was covered with thick forest. 'Bansudha', the forested area in the Ghaghra-Gomati Doab; the famous 'Namisaranya' now in Sitapur district; the hermitage of sage Valmiki on the bank of the Ganga now a part of the Unnao district; 'Khandaw' of the Mahabharata, and other such fragmentary references in old literature throw sufficient light on the predominance of natural forests in the past. During about three millennia of human occupance, gradual clearing of the natural vegetation for cultivation has taken place. Today there exist only a few small patches of woodland in the plains. More extensive forests in the submontane belt comprise hardly 4.5% of the total area of 14,422 km^2 in the region, far below the State average of 12.8%, not to speak of an optimal 33% forest cover on land. The present per capita forest area (0.001 hectare) in the region is bound to go down if the Tarai belt is excluded. Even this meagre acreage is quite unevenly distributed. The districtwise percentage of forest area is higher in the Tarai districts (Lakhimpur or Kheri 22%, Pilibhit 30.7% Saharanpur, Bijnor and Bahraich over 13% each). In the plains, it varies from 0.1% (Pratapgarh) to 3.0% (Meerut). The existing forests in the plain are of poor quality and are considered to be uneconomic.

The forests can be grouped as tropical moist deciduous, tropical wet and subtropical dry. Tropical moist deciduous forests are confined to the Tarai areas. The available species of the Tarai forest is 'Sal' but is poorer in quality as compared to the Bhabar sal. The second group of the forests is a little more open and mixed with 'Semal'. The Tarai is also known for tall grasses, like the elephant grass, 'kans', 'munj', etc. In sub-tropical dry area, especially in the western part, 'Shisham', 'Babul', 'Khair', and 'Semal' are the main species, seldom occupying large patches. The overall vegetation is made up of scattered shrubs only.

3.2.7. Ground water

The region is potentially quite rich in ground water resources, both free and confined. The confined aquifer (permanent water table) generally strikes between 60 and 90 m depth, while the temporary water table (free aquifer) depth is less than 30 m with wide spatial and seasonal variations. The free aquifer bears significant relationship with the relief, geological structure and existing water channels. Riverin 'kadar' tracts generally show lower water table, while the heavy clay belts of the 'bhangar' have higher water table.

3.2.8. Physical resource base

The total physical resource base in the region is confined to its rich alluvial soil and abundance of water resources in its long perennial rivers. The two together can serve as the best ingredients for better development of agriculture.

There are no mineral resources other than glass sands and lime concretions. The Oil and Natural Gas Commission is investigating oil resources in the 'tarai'.

3.3. THE CULTURAL SETTING

3.3.1. Population

With a population of about 69 million over an area of 150 thousand km^2, the Upper Ganga Plain is one of the most densely populated (469 persons per km^2) parts of India (Table 3.1) The large population and its spatial distribution, density and occupational structure reveal that here man has been not only the creator of cultural landscape but also a basic constituent of the ecosystem of the region. The regional distribution of population and its growth have been mainly related to the agricultural economy and public health conditions in the various parts of the region. The growth rate and density of population have been generally higher in 'bhangar' tracts of comparatively greater agricultural prosperity. In 'khadar' and 'bhur' tracts which are agriculturally less suitable, the growth and density of population have been rather low. This relation between population and agriculture was more pronounced in the earlier decades. Later, urbanization assumed greater significance and influenced the growth, distribution, density and structure of human population.

The population of the region has shown an increase of 25.85% in the growth for the decennial 1971—81. The urban population increased by 89.71%. The maximum and minimum growth rates for the urban population are in Unnao (465.69%) and Lucknow (28.56%) districts respectively (Table 3.1).

As the region is predominantly agricultural, 73.81% (59.09% cultivators and 14.72% agricultural labourers) of its total workers were engaged in this activity. The region showed a decrease of 0.85% in cultivators in 1981 compared to the ones in 1971. Also, there was a decrease of 1.02% in the agricultural labourers, whereas, workers in the household industry and other activities showed an increase of 0.28% and 1.59% respectively. Hence, a net shift of 1.87% has taken place from agricultural workers to the industrial workers in a decade 1971—81, reflecting a possible increase in the industrialization of the region (Table 3.2).

3.3.2. Urban settlement

The region may not yet be considered as an urbanized one, with 54.9 millions of the total population (around 78%) still living in the rural areas. The urban area is inhabited by 14.79 million people, which accounts for only 19.94% of the total population. The urban population is very unevenly distributed. The degree of urbanization increases from east to west in general and the existence of a few well-marked belts of urban concentration is noteworthy.

About a hundred towns consisting of about one-third of the State's urban population are studded in the western zone covering only one-seventh of the State's total area. It is the

TABLE 3.1 **Distribution of total/urban population by districts situated in the Upper Ganga Plain**

Districts	Population, 1981		Percentage of Urban to total population		Percentage growth rate 1971-81		Density of population per km²
	Total	Urban	1971	1981	Total	Urban	
Saharanpur	2,673,653	723,658	23.50	27.07	+30.12	+49.89	478
Muzaffarnagar	2,288,410	495,044	13.86	21.63	+26.97	+98.16	548
Bijnor	1,925,637	480,366	18.10	24.95	+29.22	+78.11	397
Meerut	2,766,496	865,638	22.64	31.29	+25.30	+73.20	707
Ghaziabad	1,866,778	632,166	24.63	33.86	+39.22	+91.41	721
Bulandshahr	2,349,530	460,891	13.92	19.62	+24.22	+75.01	540
Moradabad	3,151,044	850,835	23.77	27.00	+29.73	+47.39	528
Rampur	1,177,022	313,975	19.55	26.68	+30.60	+78.35	497
Budaun	1,964,094	317,743	9.35	16.18	+19.33	+106.50	380
Bareilly	2,264,770	651,142	22.28	28.75	+27.24	+64.22	550
Pilibhit	1,006,336	163,567	13.67	16.25	+33.80	+59.10	288
Shahjahanpur	1,648,659	319,650	15.24	19.39	+28.19	+63.07	360
Aligarh	2,565,450	591,337	17.85	23.05	+21.48	+56.84	511
Mathura	1,543,568	328,218	16.49	21.26	+19.63	+54.24	405
Agra	2,852,474	1,110,364	36.61	38.93	+23.56	+31.37	594

Etah	1,837,575	288,941	9.82	15.72	+16.97	+87.36	413
Mainpuri	1,724,057	191,824	8.44	11.13	+19.27	+57.20	397
Farrukhabad	2,002,513	315,014	10.91	15.73	+28.62	+85.41	469
Etawah	1,748,737	258,020	9.79	14.75	+20.79	+82.10	404
Kanpur	3,790,549	1,782,665	42.80	47.03	+26.57	+39.02	614
Fatehpur	1,572,770	141,989	5.63	9.03	+23.04	+97.46	379
Allahabad	3,780,665	769,744	18.46	20.36	+28.71	+41.99	521
Kheri	1,962,826	190,026	6.21	9.68	+32.04	+105.78	256
Sitapur	2,338,101	240,236	7.54	10.27	+24.08	+69.09	407
Hardoi	2,293,994	266,553	7.90	11.62	+24.03	+82.36	383
Unnao	1,826,463	216,067	2.57	11.83	+23.04	+465.69	401
Lucknow	2,017,172	1,058,633	50.90	52.48	+24.68	+28.56	798
Rai Bareli	1,888,181	139,833	3.40	7.41	+24.98	+172.03	410
Baharaich	2,221,154	159,503	5.93	7.18	+28.62	+55.64	323
Gonda	2,838,305	206,647	5.65	7.28	+23.30	+58.82	386
Bara Banki	2,012,576	176,079	5.76	8.75	+23.05	+86.89	457
Pratapgarh	1,807,252	91,632	1.96	5.07	+27.03	+228.32	486
Upper Ganga Plain	69,706,811	14,798,000	15.47	19.94	+25.85	+89.71	469

From : Appendix I of Paper 2 Rural-Urban Distribution and Appendix III of Paper 3 Workers/Non-Workers, Census of India, Series 1, 1981.

TABLE 3.2 **Distribution of main workers by cultivators, agricultural labourers, household industry and other workers in different districts situated in the Upper Ganga Plain**

District	Percentage of main workers in agriculture, household industry and other workers to total workers							
	Cultivators		Agricultural Labourers		Household Industry		Other workers	
	1971	1981	1971	1981	1971	1981	1971	1981
Saharanpur	36.06	34.94	26.88	28.45	4.38	3.23	32.68	33.38
Muzaffarnagar	43.23	41.38	27.97	28.64	5.48	4.13	23.32	25.85
Bijnor	45.61	42.55	23.20	23.40	7.02	8.28	24.17	25.77
Ghaziabad	38.22	32.94	12.61	12.78	6.46	4.40	42.71	49.88
Meerut	38.95	37.04	16.32	18.94	7.17	6.50	37.56	37.52
Bulandshahr	54.44	52.33	17.51	18.09	4.66	4.02	23.39	25.56
Moradabad	62.68	59.33	10.11	10.35	4.32	4.91	22.89	25.41
Rampur	67.50	64.21	9.33	10.60	1.62	2.79	21.55	22.40
Budaun	79.83	78.47	8.37	7.99	1.50	1.62	10.30	11.92
Bareilly	62.86	59.02	10.56	11.67	2.11	2.40	24.47	26.91
Pilibhit	68.22	67.10	13.61	13.66	1.82	1.69	16.35	17.55
Shahjahanpur	69.18	69.81	13.68	11.11	1.70	2.01	15.44	17.07
Aligarh	49.01	44.70	18.97	20.07	4.23	6.26	27.79	28.97
Mathura	53.59	53.08	15.19	15.36	4.79	4.03	26.43	27.53

Agra	41.74	39.69	10.41	10.02	5.43	8.92	42.42	41.37
Etah	69.89	70.28	12.38	11.83	2.62	2.77	15.11	15.12
Mainpuri	70.42	70.88	11.46	10.85	3.00	2.07	15.12	16.20
Farrukhabad	68.69	67.26	11.68	10.09	3.00	3.43	16.63	19.22
Etawah	67.99	66.80	13.07	12.65	2.55	2.56	16.39	17.99
Fatehpur	59.20	61.38	27.11	22.66	2.92	2.53	10.77	13.43
Allahabad	44.54	46.84	28.38	23.55	4.75	6.56	22.33	23.05
Kheri	76.33	77.49	12.27	9.96	1.04	1.45	10.36	11.10
Sitapur	77.51	72.70	9.72	8.13	2.87	3.23	9.90	15.94
Hardoi	77.01	79.13	11.25	8.20	2.32	1.95	9.42	10.72
Unnao	72.33	74.39	14.18	10.22	2.80	2.80	10.69	12.59
Lucknow	38.61	38.21	7.76	7.39	3.62	7.38	50.01	47.02
Rae Bareli	65.49	67.08	22.40	17.87	2.80	2.37	9.31	12.68
Bahraich	76.85	79.59	14.11	10.07	0.96	1.27	8.08	9.07
Gonda	70.88	76.36	18.88	12.96	1.81	1.53	8.43	9.15
Bara Banki	71.80	72.63	15.36	12.13	4.34	4.66	8.50	10 58
Pratapgarh	61.63	55.65	5.96	28.84	3.36	2.29	9.05	13.22
Upper Ganga Plain	59.94	59.09	15.74	14.72	3.46	3.74	20.86	22.45

From : Appendix II, Paper 3 of 1981, Workers and Non-workers, Census of India, Series 1, 1981.

most urbanized part of the region. The urban concentration is heaviest in the Doab and adjoining regions where towns are close to each other and form rather a continuous zone. In the other parts of the region, the urban centres are generally scattered. Even the two largest towns of the region, Kanpur and Lucknow, stand isolated.

The western part of the region has two distinct urban belts : (i) In the Doab from Saharanpur in the north to Etawah in the south, with two conspicuous agglomerations—(a) Aligarh —Hathras—Mathura—Agra—Firozabad, and (b) Ghaziabad—Hapur—Meerut—Muzaffarnagar—Saharanpur; (ii) Through Rohilkhand from Bijnor to Fatehpur comprising (a) Amroha—Moradabad, (b) Rampur, (c) Pilibhit—Bareilly—Badaun, and (d) Etah.

The region contains most of class I cities of Uttar Pradesh (22 out of 30) of which 8 cities are situated along or in the vicinity of the Ganga river. The total population in 1981, growth rate for the two decennials 1961-71 and 1971-81, total main workers and the distribution of these workers in different categories, for these 8 cities are given in Tables 3.3 and 3.4.

3.3.3. Agricultural economy

The regional economy is dominated by agriculture which together with the allied activities forms the most important source of employment and revenue. Though, in general, the proportion of the net sown area to the total area is high, being over 65%, there is a wide sub-regional variation which gets sharper at district and tahsil levels.

The region has very low percentage of land under forests (8.4) as compared to U.P. as a whole (12.8) and the percentage ranges between almost nothing (Pratapgarh) to 30.7 (Pilibhit). Most of the plain's districts have negligible forest cover, as many as 24 districts having below 5% and of them 7 districts recording even less than 1%.

The situation of severe concern is obtained however, in case of permanent pasture and other grazing lands which are so important in a region of a high density of bovine population.

TABLE 3.3 **Population of Class I cities situated along or in the vicinity of the Ganga river in the Upper Ganga Plain**

Name of City/Urban agglomeration (U.A.)	Population	Growth rate 1961-71	Growth rate 1971-81
Kanpur (U.A.)	1,688,242	+31.32	+32.39
Allahabad (U.A.)	642,420	+19.11	+25.22
Meerut (U.A.)	538,461	+29.49	+46.42
Aligarh (M.B.)	319,981	+36.37	+26.82
Farrukhabad-cum-Fatehgarh (U.A.)	160,927	+17.17	+45.20
Etawah (M.B)	112,426	+23.27	+30.89
Bulandshahr (M.B.)	103,666	+34.74	+74.21
Hapur (M.B.)	103,466	+28.99	+45.18

M.B.=Municipal Board

From : Appendix II, Paper 2 of 1981 Rural-Urban Distribution, Census of India, Series 1, 1981.

The regional average is below 0.2%, though slightly above that in the State (below 0.15%). Land put to non-agricultural uses varies from 5.0% (Mathura) to 10.6% (Saharanpur) of the total. Land under miscellaneous trees and groves is high in the rather compact blocks in the districts like Faizabad (7.0%), Sultanpur (8.3%) and Unnao (5.1%).

There is appreciable variation in the use of the agricultural land. As many as 21 districts have higher percentage of agricultural land under net sown area than U.P. as a whole (84.3%); as such most of them have lower percentage under cultivable waste and current and other fallows than the State averages (8.3 and 7.2% respectively). 5 to 10% of the total agricultural area remains under current or old fallows, which could be cropped with proper soil management, availability of irrigational water and use of manure and fertilizers (Singh, 1971). Recent data about land use pattern in certain districts is given in Table 3.5.

TABLE 3.4 **Distribution of population by workers and non workers in the Class I cities situated along or in the vicinity of the Ganga river in the Upper Ganga Plain**

Urban agglomeration (U.A.)/city	Total main workers	Workers as per cent to total	Cultivators	Agricultural labourers	Household industry	Other workers	Non-workers
Kanpur (U.A.)	454,749	26.94	6779	4717	46,882	4,10,028	1,219,836
Allahabad (U.A.)	161,436	25.13	4075	6694	14,839	1,65,093	451,719
Meerut (U.A.)	153,024	28.42	2401	3831	15,405	1,31,597	385,227
Aligarh (M.B.)	81,346	25.42	1743	1256	12,681	66,390	237,911
Farrukhabad-cum-Fatehgarh (U.A.)	47,405	29.46	2573	2359	3,460	39,150	113,385
Etawah (M.B.)	28,852	25.66	740	444	4,242	23,569	83,431
Bulandshahr (M.B.)	25,767	24.86	941	579	1,487	22,850	77,809
Hapur (M.B.)	26,729	25.83	1018	1655	1,151	22,927	76,715

U.A.=Urban Agglomeration, M.B.=Municipal Board
From : Appendix IV, Paper 3 of 1981, Workers and Non-workers, Census of India, Series 1, 1981.

TABLE 3.5 **Land use pattern in different administrative divisions situated in the Upper Ganga Plain (Data pertains to the years 1971-72)**

Administrative Divisions	Total area	Crops		Forests		Cultivable land		Uncultivable land & non-agricultural area	
	in '000 of ha	'000 of ha	%	'000 of ha	%	'000 of ha	%	'000 of ha	%
1	2	3	4	5	6	7	8	9	10
Agra	485	356	73	40	8	38	8	52	11
Aligarh	503	388	77	1	—	40	8	73	14
Allahabad	742	479	64	20	3	130	17	113	15
Bahraich	691	449	65	101	15	75	11	66	9
Banda	791	493	62	71	9	144	18	82	10
Bara Banki	448	304	68	3	—	79	18	62	14
Budaun	522	406	78	7	1	55	10	54	10
Etah	445	310	69	1	—	87	19	47	11
Etawah	446	292	65	40	9	44	10	70	16
Farrukhabad	432	290	67	5	1	72	17	64	15
Fatehpur	427	293	68	7	2	59	14	68	16
Faizabad	437	305	70	1	—	70	16	60	14
Gonda	746	522	70	72	10	80	11	71	9
Hamirpur	728	506	70	38	5	114	—	69	9
Hardoi	605	389	64	10	2	130	21	75	12

TABLE 3.5 (*Contd.*)

1	2	3	4	5	6	7	8	9	10
Jalaun	460	360	78	26	6	31	7	42	9
Jhansi	1016	475	47	96	9	345	34	99	10
Kanpur	620	432	70	12	2	69	11	107	17
Kheri	795	439	55	175	22	104	13	78	10
Lucknow	256	152	59	11	4	51	20	40	16
Mathura	373	309	83	2	—	30	8	32	9
Mainpuri	436	282	65	7	1	56	13	91	21
Pratapgarh	368	253	69	—	—	57	15	57	16
Rae Bareli	458	283	62	5	1	93	20	77	17
Shahjahanpur	461	308	67	13	3	93	20	46	10
Sitapur	573	424	74	6	1	82	14	61	11
Sultanpur	445	312	70	1	—	64	14	67	15
Unnao	469	293	62	16	3	87	19	72	15

From : Meher Homji *et al.*, 1978.

3.3.3.1. Irrigation

Irrigation has played a dominant role in boosting the agricultural prosperity of this region, particularly during the last one hundred years. The Upper Ganga Plain is one of the highly irrigated agricultural regions of India, where 30% of the net sown area is irrigated. There is a wide sub-regional and district level variation in the irrigated percentages. Most of the Upper and Middle Doab, down to Farrukhabad and Etawah, have greater proportion of irrigated areas : Muzaffarnagar (59.1%), Meerut (68.6%), Bulandshahr (61.2%) in the Upper Doab record the highest percentages, more than double the State average, while most of the other districts also in these parts, have 30 to more than 50% of their net sown area under irrigation. In the Lower Doab, however, irrigation becomes less important as is evidenced by the lower percentages in Kanpur (29.3), Fatehpur (26.2) and Allahabad (23.2) districts. Rohilkhand and Awadh, on the other hand, range between very poor to poor in irrigational facility. The districts from Bijnor (12.6) to Bahraich (7.9), irrigate less than 15% of their net sown area, and in the entire sweep of these vast tracts, only Moradabad (22.3) and Budaun (18.5) in the south-west Rohilkhand, and Unnao (28.3), Lucknow (27.7), Bara Banki (28.4) and Gonda (26.0) in the east, have 15 to 30% of the net sown area irrigated. The 'tarai' districts, which have higher rainfall and lower temperature range and evapotranspiration rate, have low irrigational needs. Hence Pilibhit (11.1), Kheri (5.6) and Bahraich, have the lowest percentages of irrigated land in U.P. The eastern fringe of the Awadh Plain is better irrigated as is evidenced by the district averages of Faizabad (49.3), Rae Bareli (40.5), Sultanpur (39.6) and Pratapgarh (38.7).

Although surface wells, being cheap and handy, are the ubiquitous source of irrigation throughout the region, there are wide variations in the levels of dominance of various sources of irrigation in different sub-regions. In general, wells including tube wells (47.2%) and canals (39.5%) are by far the most important in the State, and unlike in the Middle Ganga Plain, tanks, 'jhils', 'tals' and other sources (13.3%) are less important here (from Singh, 1971).

The Upper Ganga Plain has extensive canal network developed meticulously over the last hundred years. The important systems are the East Yamuna, the Upper Ganga and the Lower Ganga Canals.

The Upper Ganga Canal (1853) takes out from the right bank of the Ganga at Hardwar where a dam is built across the river. The main channel and the principal branches have a total run of about 900 km irrigating over 2.4 million acres, mainly in Saharanpur, Muzaffarnagar, Meerut and Bulandshahr districts down to Mainpuri. The Deoband and Anupshahr branches are important. A 36 m fall in the bed-slope of the canal from Hardwar to Aligarh is broken into 13 vertical falls of 2.4 to 3.6 m, seven of which have been utilized for generating electricity for about 2,000 tubewells under the Ganga Valley Tubewell Scheme. As mentioned above, these tubewells provide water to the 'bhur' tracts of Bijnor and Moradabad and contiguous areas in western Rohilkhand, which are not amenable to canal irrigation.

The Lower Ganga Canal (1872) takes off on the right bank of the Ganga at Naraura (Bulandshahr) by constructing a dam across the river. The main branches run for over 1,000 km, while the distributaries are over 5,000 km. The system irrigates over one and a half million acres of Central and of Lower Doab in Aligarh, Etah, Etawah, Mainpuri, Farrukhabad, Kanpur, Fatehpur and right down to Allahabad districts. The other canals in the region are the Sarda canal, the Ramganga Multipurpose Project, and the Agra canal in the Yamunapar tract.

As noted above, as many as seven falls in the bed-slope of the Upper Ganga Canal have been utilized as sites for generating electricity at Bahadurabad, Miragajni, Chitaura, Salawa, Bhola, Palra and Sumera. The transmission system of high tension lines runs for over 3,000 km in length and provides power for irrigational, residential and industrial purposes in the area.

3.3.3.2. Cropping pattern

The region is more diversified in cropping than the Middle Ganga Plain. There is a very high premium on food-crops, and over 85% of the total cropped area is devoted to such crops. Cereals alone claim over 3/5th of the cropped land and pulses about 1/5th, while about 8% is devoted to sugarcane, a crop which has almost replaced cotton in many districts reducing the latter to a mere 0.4% in the region. Oilseeds are next to sugarcane as cash crops and are of increasing importance in some tracts.

Wheat is the premier crop of the region, which places Uttar Pradesh in the first position by producing about 35% of the country's total. Over 1/5th of the total cropped land is devoted to wheat in the region as a whole but the districts of the Upper Doab and contiguous parts of the Western Rohilkhand Plain provide more than 1/4th of the cropped land to this prized crop. In fact, the cultivation of wheat replaces rice as one goes west where the fertile loamy soil, adequate and assured irrigation, the small but highly beneficial winter rain and suitable temperature regimes are significant factors for its regional dominance.

Rice, though the first crop in U.P., is second to wheat in this region and claims over 1/6th of the total cropped land. It is, however, an important crop in the humid sub-Himalayan 'tarai', e g., Saharanpur, Bahraich and Pilibhit merging with the Eastern Awadh and lower Doab belt. The Central Awadh tract (Hardoi to Lucknow), Central Doab (Etah to Etawah) and south-west Rohilkhand (Budaun to Moradabad) also devote sizable portion of the crop land to rice, but in the Upper Doab, it is rather unimportant.

Millets and maize together, as in the State itself, occupy third place in the region, claiming 16.8% of the total cropped land. Being hardier crops, millets gain importance in the drier west and south-west, particularly in the saline and alkaline soil zones with less than 100 cm rainfall. Bajra is the most important millet claiming over 7% of the total crop land, and is particularly so in the south-western plain. Maize covers over 6% of the crop land and forms a belt in the tarai region. Jowar gets less than 4% of the cropland.

Pulses and oilseeds are grown throughout, as in U.P. as a whole, mostly interculturally. Several kharif (urd, moong, arhar and moth) and rabi (gram, peas and masur) crops consist of pulses which provide greater diversity to the cropping pattern of the region. Gram (12.0) and peas (4.6) together take about 1/6th of the total cropped land in the region and are close to rice in acreage. Arhar is significant in Agra and Kanpur but is negligible in the 'tarai' districts. Masur, on the other hand, attains prominence in the 'tarai' districts from Bareilly to Gonda and in Shahjahanpur; Bahraich being at the top.

Among the several oilseeds, groundnut is the most important and is rather concentrated in a broad northwest-southwest belt from Moradabad district to Lucknow in the Upper Central part of the region. As it thrives well on light sandy soil not suitable for most other crops, it assumes great importance in the 'bhur' tracts of Moradabad and Budaun and the sandy soils of Sitapur and Hardoi, which together account for about half of the groundnut area of the State. Mustard, next to groundnut in acreage, is more widespread, but forms two well-marked clusters, one from Mathura to Kanpur in the south and beyond the region in Jalaun and the other in the Tarai districts of Kheri, Bahraich, Gonda and further east to Gorakhpur and Kanpur. Bahraich and Mathura are most important districts in order of acreage. Allahabad is important for linseed.

Sugarcane, cotton and jute are the commercial crops of the region. While cotton is fast decreasing in importance jute, a rather new entrant, is gaining ground. U.P. grows about 46% of the country's sugarcane, and the Upper Ganga Plain provides the bulk of it with 7.5% of the cropland under sugarcane. As it requires deep fertile loamy soil and plenty of assured water supply, it forms an east-west belt in the north of the plain with important sub-regional clustering in the (i) Upper Doab and Western Rohilkhand, and (ii) Central Tarai and neighbouring districts. There has been an extensive increase in sugarcane acreage during the past four decades. The highest productivity accounting for more than 1/4th of the total value of agricultural produce is recorded in this region. Cotton is concentrated in the West Plain between Muzaffarnagar and Mathura. Jute is confined to the more humid 'Tarai' districts of the regions, particularly in Kheri and Bahraich and further in the East U.P. plain. Area under different crops in some of the administrative divisions lying in the Upper Ganga Plain is given in Tables 3.6 and 3.7 (Fig. 3.3).

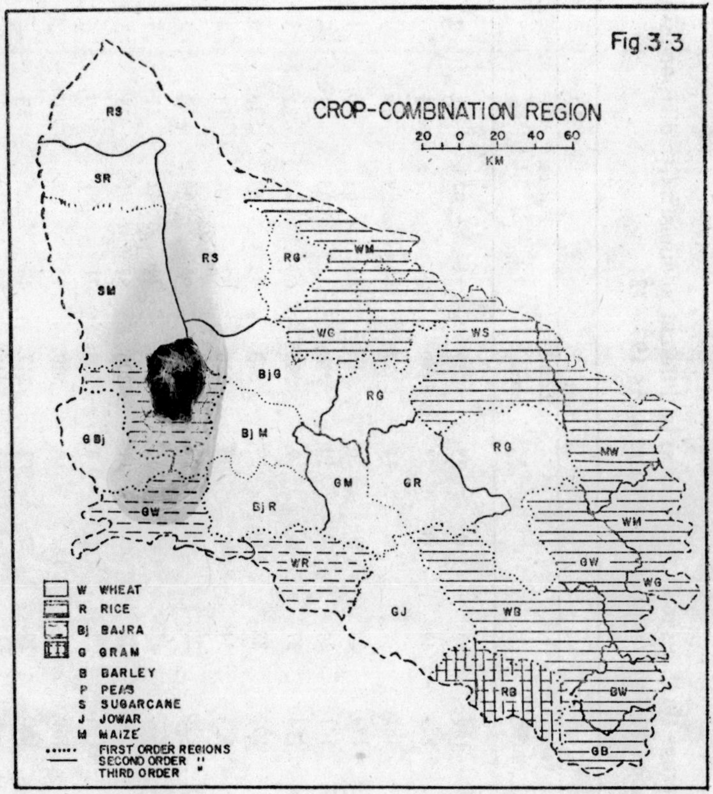

Fig.3·3

3.3.4. Fisheries

The Ganga system which has a combined length of about 8,047 km and is regarded as one of the largest systems of the world, comprises the icy cold Himalayan streams and the warm, biologically more productive, waters of the North Indian plains. This system harbours the richest freshwater fish fauna of India ranging from mahseers and the torrential fishes of the hills to the culturable Gangetic carps, the hilsa yielding bumper catches in certain years of profusion, and a wide range of other fishes, some of great commercial value.

TABLE 3.6 **Area under food crops (in thousands of hectares) in different administrative divisions situated in the Upper Ganga Plain (Data pertains to the years 1971-72)**

Administrative Divisions	Food crops											
	Paddy	%	Wheat & barley	%	Millets	%	Total cereals	%	Pulses	%	Potato	%
Agra	2	—	147	33	110	25	259	58	97	22	2	—
Aligarh	12	2	247	42	190	32	449	76	67	11	3	1
Allahabad	150	24	181	29	66	11	397	64	128	20	9	1
Bahraich	181	29	174	28	161	26	516	83	64	10	1	—
Banda	84	14	141	24	74	13	299	51	243	41	—	—
Bara Banki	130	30	114	27	39	9	283	66	90	21	5	1
Budaun	37	7	197	38	139	27	373	72	67	13	5	1
Etah	22	5	155	34	149	33	326	72	58	18	4	1
Etawah	67	17	110	28	91	23	[illegible]	68	64	16	5	1
Farrukhabad	28	7	122	31	141	36	[illegible]	74	42	11	24	6
Fatehpur	69	19	96	27	41	11	206	57	91	25	2	1
Faizabad	152	36	116	27	32	8	300	71	81	19	6	1
Gonda	249	34	149	21	131	18	529	73	67	9	2	—
Hamirpur	10	2	150	29	87	17	247	48	227	43	—	—
Hardoi	62	12	188	36	69	13	319	61	99	19	5	1

Jalaun	15	4	136	36	40	10	191	50	154	40	—	—
Jhansi	17	3	176	33	118	22	311	58	147	28	1	—
Kanpur	79	15	163	31	95	18	337	64	116	22	6	1
Kheri	130	23	149	26	52	9	331	58	68	12	2	—
Lucknow	42	22	61	32	20	10	123	64	32	17	4	2
Mathura	5	1	188	46	90	22	283	69	57	14	1	—
Mainpuri	39	10	137	35	97	25	273	70	40	10	7	2
Pratapgarh	87	28	81	26	39	13	207	67	66	21	4	1
Rae Bareli	118	31	122	32	38	10	278	73	72	19	2	1
Shahjahanpur	96	23	131	31	43	10	270	64	69	17	3	1
Sitapur	97	18	177	33	69	13	343	64	90	17	2	—
Sultanpur	151	37	95	23	50	12	296	72	88	22	4	1
Unnao	79	20	152	38	47	12	278	70	61	15	3	1

From : Meher Homji *et al.*, 1978.

TABLE 3.7 **Area under industrial crops (in thousands of hectares) in different administrative divisions situated in the Upper Ganga Plain (Data pertains to the year 1971-72)**

Administrative Divisions	Sugarcane	%	Tobacco	%	Oilseeds	%	Fibres	%
Agra	2	—	—	—	45	10	2	—
Aligarh	12	2	—	—	4	1	13	2
Allahabad	5	1	—	—	13	2	3	—
Bahraich	6	1	—	—	23	4	2	—
Banda	1	—	—	—	10	2	2	—
Bara Banki	19	4	—	—	4	1	—	—
Budaun	19	4	—	—	55	11	—	—
Etah	8	2	2	—	21	5	2	—
Etawah	5	1	—	—	26	7	1	—
Farrukhabad	11	3	2	1	25	6	—	—
Fatehpur	8	2	—	—	5	1	2	1
Faizabad	26	6	—	—	1	—	1	—
Gonda	25	3	—	—	12	2	1	—
Hamirpur	2	—	—	—	23	4	3	1
Hardoi	18	3	—	—	57	11	1	—

Jalaun	2	1	—	—	14	4	1	—
Jhansi	1	—	—	—	16	3	1	—
Kanpur	6	1	—	—	34	7	2	—
Kheri	77	14	—	—	31	5	5	1
Lucknow	2	1	—	—	5	3	—	—
Mathura	13	3	—	—	7	2	8	2
Mainpuri	2	1	—	—	20	5	—	—
Pratapgarh	4	1	—	—	1	—	5	2
Rae Bareli	3	1	—	—	7	2	—	—
Shahjahanpur	26	6	—	—	16	4	1	—
Sitapur	37	7	—	—	33	6	3	1
Sultanpur	12	3	—	—	—	—	2	1
Unnao	8	2	—	—	19	5	1	—

From : Meher Homji *et al.*, 1978.

Detailed investigations on the fisheries of the Ganga river system have been conducted during the years 1958-66 by the Allahabad Substation of the Central Inland Fisheries Research Institute (CIFRI) in the stretch lying between Agra in the west (78° E) and Rajmahal in the east (88° E). The investigations comprised total enumeration of daily arrivals of fish at Agra and Allahabad on the river Yamuna, and at Kanpur, Varanasi, Buxar, Ballia, Patna and Bhagalpur on the river Ganga. In this chapter, however, the data for Allahabad and Kanpur only have been incorporated. The year was reckoned from March to February (Table 3.8) (From Jhingran, 1977).

TABLE 3.8 **Fish landings (tonnes) from Kanpur and Allahabad Centres situated along the Ganga river system in the Upper Ganga Plain during the period 1955-56 to 1965-66**

Year	Kanpur	Allahabad
1955-56	—	316.42
1956-57	—	485.00
1957-58	—	200.06
1958-59	93.35	145.35
1959-60	105.62	193.58
1960-61	81.76	200.11
1961-62	29.71	202.82
1962-63	52.43	263.08
1963-64	43.09	200.41
1964-65	33.72	256.35
1965-66	32.62	212.49
Average	59.02	243.25

From : Jhingran *et al.*, 1970.

3.3.5. Industrial economy

The Upper Ganga Plain, though industrially not very important in the country, is comparatively more industrialized than the Middle Ganga Plain. As per 1961 data (From Singh, 1971) out of the 16 districts in U.P. with higher proportion of industrial workers than the State average (9.1) as many as 12 districts fall in this region in a block from Saharanpur to Agra and Bijnor, Moradabad and Bareilly in the west and Kanpur and Lucknow districts in the east. Districts of Meerut with 22.5% industrial workers, Agra with 19.3% and Kanpur with 18.7% lead the list. Other districts in the Central and Lower Doab and most of Rohilkhand and Awadh Plain have between 5 and 10% industrial workers (Fig. 3.4). Household industries have a greater importance everywhere.

Uttar Pradesh is by far the largest single producer of sugar in the country, and the bulk of the production comes from the Upper Ganga Plain. According to 1961 data (From Singh, 1971), of the State's 71 factories employing about 55,000 workers, 43 were in this region with about 40,000 workers. The factories in this region are, in general bigger in size and as many as 14 units have more than 1,000 workers each. The profitability index is also high in this region. They also economize more by developing by-product industries. Most of the distilleries are attached to the sugar mills, using their molasses. The largest distillery is, however, outside the sugarbelt at Lucknow. There is one wine-making unit at Rampur and one of soft drinks at Kanpur.

Other food processing industries include rice, dal and edible oil mills, bakeries, confectionery, fruit processing, dairy and hydrogenated oil units. They are mostly very small in size and are located both in large and small centres, mostly in Doab with Kanpur, Allahabad, Etawah, Aligarh, Hathras, Ghaziabad and Meerut as the important centres. Outside this region Lucknow, Bahraich, Shahjahanpur, Bareilly and Rampur are also notable. Peculiarly enough, vast majority of U.P. rice mills are located in this region instead of in the rice dominated East U.P. Plain. Most of the rice mills are centred in small market towns. Similarly most of the State's dal mills are located in this region. Aligarh district has more than 1/3rd of the State's dal mills. Kanpur leads in the edible oil industry, other centres are Agra, Aligarh and Hathras. All the factories of U P. producing hydrogenated oils are located in this region, viz., Ghaziabad, Modinagar and Kanpur. Bakery and sugar confectionery are centred mainly at Kanpur, Aligarh and Saharanpur being other centres. Kanpur enjoys the premier position having factories of all categories.

Textile industries comprising cotton, wool, jute, silk and others are the largest single industrial group, both in terms of employment and industrial output in the region as also in U.P. as a whole. Cotton group is by far the largest. Kanpur which is advantageously located between the Bihar which supplies coal and Punjab Plain and Malwa Plateau supplying cotton, has more than a dozen large cotton factories, employing two-third of the State's cotton workers. Kanpur's industrial growth has been sudden and fast. The first cotton mills were established in the city between 1869 and 1882 after it had become a great cotton trading centre. American Civil War provided stimulus for the Indian cotton trade. The city specializes in the manufacture of tent canvas and other types of cloth. Modinagar, Hathras, Rampur, Saharanpur, Agra and Allahabad are other important cotton textile factories, 8 are located in this region, and Kanpur is the leading centre with about two-thirds of the State's total workers in this group in its two large factories. There is only one small silk textile unit at Kanpur, most others being at Varanasi and Dehra Dun outside this region. Kanpur, however, is the largest knitting centre, employing over 65% of workers in the State in this group. The city also has two units producing wearing apparel, the only units in the State.

Kanpur and Agra are world famous leather and leather goods centres and the former, employs about four-fifths of the total workers engaged in the State, in this group. Majority of the tanneries are located at Kanpur employing over 90% of the State's workers in the region. Agra, Unnao, Farrukhabad and Fatehpur are other tannery centres. Meerut is also an important leather goods centre.

All the paper and pulp industry units of U.P. are located in this region. Saharanpur and Lucknow paper mills employ about 60% of the total workers in the State in this group. Paper is also manufactured at Pilkhuwa (Meerut) while Meerut, Saharanpur, Kanpur and Allahabad prepare paper-board and straw board. Saw milling units are concentrated in Lucknow district, other notable centres are Clutterbuckganj (Bareilly), Saharanpur and Haldwani (Naini Tal). Sitapur plywood unit is the largest of the State's three units, others being at Najibabad (Bijnor) and Jwalapur (Saharanpur). Two government-owned wood working institutes at Bareilly and Allahabad engage about 55% of the total workers in the manufacture of wood products in the State.

As the region like most of the State lacks in mineral resources of any significance, no heavy metallurgical or mineral based industry has developed, excepting some units using imported materials. According to 1961 statistics, out of over 350 units of metal based industries more than 300 are located in the Upper Ganga Plain (Singh, 1971). Basic ferrous metal industries form the largest group, employing about half of the total workers, and Kanpur is the chief centre engaging half of the State's workers in this group.

Kanpur, Bareilly, Lucknow, Rampur and Ghaziabad manufactured agricultural implements, while Agra and Kanpur turn out machine tools, wood working machinery and other tools. Kanpur also manufactures textile machinery and accessories and a few other machines. General and job engineering industry is more widespread, the important centres being Lucknow, Kanpur, Meerut and Ghaziabad with about 60% of the workers in the industry. Kanpur, Bulandshahr and Ghaziabad prepare fans, radiators and other electrical goods and accessories. Insulated wires and cables are made at factories in Agra, Ghaziabad, Mathura, Aligarh and Kanpur. Kanpur has about half of U.P.'s bicycle manufacturing units, though Meerut and Ghaziabad units are the biggest. Kanpur has become a large aeronautical engineering centre. Aeroplanes are repaired at Lucknow and at Allahabad.

All the factories of U.P. producing heavy chemicals are located in this region, Kanpur having the largest number. Fertilizers are produced at Kanpur, Moradabad, Allahabad and at some other smaller centres. About half of the State's fine and pharmaceutical chemical units are concentrated in Lucknow and Kanpur. Meerut, Agra and Kanpur manufacture rubber goods.

The region is famous for cottage and small scale industries. Its handicrafts have been known in the international market since ancient times. Handloom weaving is the most important cottage industry. It is widely scattered, furnishing from very coarse country cloth on simple handlooms, such as garha and gazi, to highly artistic brocades, chikans, embroideries and fine silk and cotton fabrics. Etawah and Sandila are famous for silk and art silk weaving. Wool weaving, especially blankets, is important in the district of Muzaffarnagar, Meerut and Bijnor (Najibabad). Calico-printing, though widespread, is especially important at Farrukhabad, Lucknow, Mathura, Pilkhuwa, Amroha and Rampur. Other industries include leather tanning and manufacturing leather goods, glue and gut industry, etc. The sports goods industry has taken roots at Meerut and Agra. Brass and copper-ware industry is traditionally centred at

Moradabad, Farrukhabad, Hathras, Shamli, Aligarh, etc. Iron and Steel goods and utensils are manufactured at Etah, Moradabad, Meerut, Etawah, Agra, Fatehpur, Hathras, Ghaziabad, Hardoi, Lucknow, Bareilly, etc. Firozabad is an important glass and glass products centre. Ceramics and pottery of Khurja and clay modelling industries of Allahabad and Lucknow are famous.

3.3.6. Transport and communication

Though the present system of the regional transport comprising roads, railways, and airways is the product of last hundred years or so, some crude and rather urbanized roads, cart-tracks and river routes have been instrumental in the process of regional land occupance and resource uses since ancient times. In the late medieval period, many roads were constructed and used for military and trade goods movements. Grand Trunk road has been important since ancient times, and so has been the Ganga system of rivers. Roads and rivers connected the region with different parts of the country including the coastal parts.

The overall regional road and railway net is shown in Fig. 3.5. The region is mostly served by Northern and North-Eastern railway systems. Because of its position, extent and high population density, the region has all the types of traffic—local, originating, terminating and bridge traffic. The main lines carry mostly terminating and bridge traffic, linking as they do in the Western and Eastern India. The most important pattern of rail traffic flow is of inter-State character. SE-NW flows are prominent and there is little transverse (N-S) movement. The density of railways in U.P. works out to 297 km per 10,000 km^2 area, in view of which the central east-west lying block of the region of the State as a whole is one of the best developed regions of India in rail transport. In this part the density exceeds 300 km everywhere. High density of over 500 km is found in Aligarh-Agra and Kanpur-Lucknow areas; Lucknow region has the highest density of over 600 km in the State. Density decreases both towards north and south because of the unsuitability of the terrain.

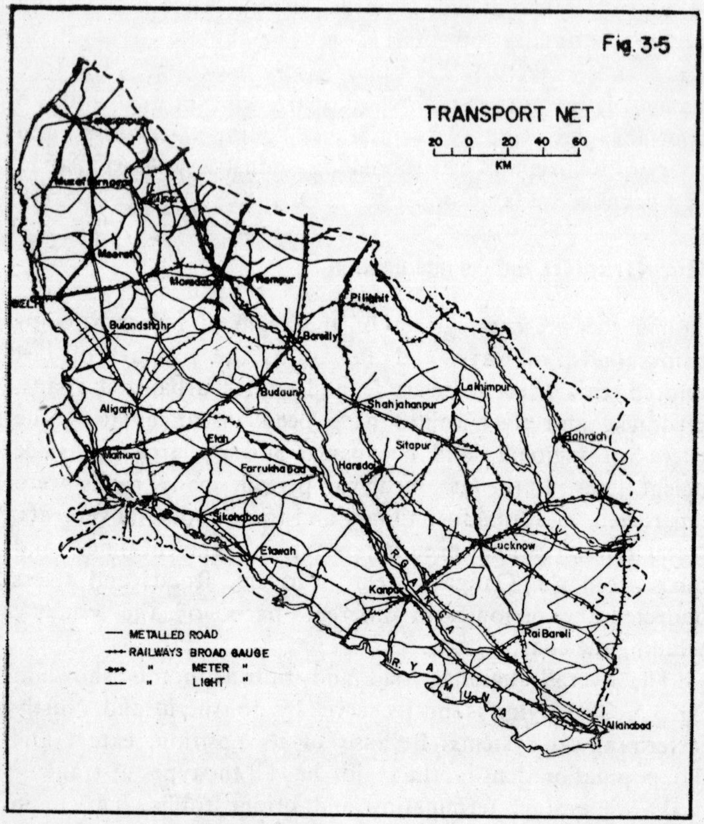

The region is inadequately served by roads as compared to several other parts of the country but is better served than the Middle Ganga Plain. Most of the region is served with 750 km of surfaced roads per 10,000 km^2 of area. The entire area from Himalayan district of Dehra Dun southward to Agra comes under a relatively high density category (1,000-1,250 km). Lucknow district enjoys the highest density (over 1,500 km) of roads. Most of the 'Tarai' and the contiguous parts of Sitapur, Hardoi, Shahjahanpur, etc., fall below 750 km decreasing towards the north to below 500 km. The marshy nature of the 'Tarai' and its slower pace of development are responsible for low density.

There is enormously heavy traffic density on trunk roads and on such other roads which run parallel to the railways. The Upper Doab region has the highest density of both passenger and freight traffic. In this entire region north to south flow is predominant.

The principal cities like Kanpur, Lucknow, Agra and Allahabad are on the internal air routes of the country.

3.4 PROBLEMS IN THE REGION

The Upper Ganga Plain is relatively more developed than the Middle Ganga Plain, and has its own environmental problems. For instance, overgrazing and deforestation in most areas have led to soil erosion and ravine formation on the one hand and have accentuated flooding on the other. The traditional pastoralism of the Braj Plain, a critical rainfall zone, could not maintain the ecosystem partly due to overgrazing and partly due to the encroachment of desertic conditions. Likewise, the vigorous activity in recent years in the use of the Tarai resources, has accentuated the processes of flood, waterlogging and soil erosion leading to imbalances in the ecosystem.

Due to an increase in exploitation of forests in the zone above the Upper Ganga Plain following increased human population, there is a recurrence of floods in the regions situated downstream. Every year there is an increase in the damages caused by these recurring floods which have led to large scale destruction of property, crops, public utilities, etc., running into several lakhs of rupees. Flood damages for the State of U.P. comprising zone above the Upper Ganga Plain and Upper Ganga Plain are given in Tables 3.9 and 3.10.

Besides floods the region also faces the grave danger of water pollution of the Ganga river, due to increased urbanization though the region itself is still mainly rural. However, due to the industrial development of certain cities at a faster rate, the pollution problems of local origin keep on occurring which are sometimes very acute in nature. It is the experience from other areas of the country that in the presence of numerous industries in limited area, industrial pollution of a river is acute (Chaphekar, 1979). In Kanpur, Lucknow and Allahabad

TABLE 3.9 **Flood damage during 1953 to 1981 in the State of Uttar Pradesh**

Sl. No.	Year	Area affected in lakh ha	Population affected in lakhs	Damage to area in lakh ha	Damage to no. of houses	Cattle lost	Human lives lost
1.	1953	9.1	143.7	NA	80443	341	23
2.	1954	13.5	34.0	7.4	19550	1555	27
3.	1955	41.3	140.0	25.7	768264	5694	231
4.	1956	25.0	86.0	NA	401889	2595	155
5.	1957	17.0	41.9	NA	263151	1281	193
6.	1958	21.8	36.6	NA	130097	1931	143
7.	1959	0.7	2.2	Neg.	116	NA	1
8.	1960	32.5	NA	5.9	374897	1417	229
9.	1961	22.9	NA	2.2	316560	1470	334
10.	1962	10.6	17.9	10.6	65315	149	28
11.	1963	12.6	63.8	12.0	313231	2013	292
12.	1964	10.6	43.0	7.7	39648	87	69
13.	1965	0.8	3.3	Neg.	1136	6	4
14.	1966	5.2	13.6	2.8	42712	161	40
15.	1967	34.3	87.6	15.2	328013	1677	226
16.	1968	8.0	18.2	4.0	19577	200	66

17.	1969	24.0	57.4	13.1	149204	3187	113
18.	1970	29.1	108.0	17.0	553000	3913	344
19.	1971	52.6	208.1	26.5	744251	3064	605
20.	1972	3.1	21.0	1.8	11822	1052	69
21.	1973	26.3	128.4	15.0	193362	269	192
22.	1974	13.5	52.2	8.0	85244	23	37
23.	1975	21.8	62.8	13.8	106679	473	148
24.	1976	36.3	141.5	21.6	290757	2425	257
25.	1977	13.0	37.8	6.1	50648	887	157
26.	1978	73.4	225.9	39.3	1197552	7430	739
27.	1979	7.0	21.1	5.2	22958	222	77
28.	1980	58.6	303.5	30.9	1923134	5244	1309
29.	1981	29.9	146.3	16.4	491102	1356	427
	Total	654.5	2245.8	309.2	8984302	50122	6535
	Maximum	73.4 (1978)	303.6 (1980)	30.3 (1978)	1923134 (1980)	7430 (1978)	1309 (1980)
	Average	22.6	83.2	12.4	309804	1790	225

NA=Not Available, Neg.=Negligible.

From : Data made available by Central Water Commission, New Delhi.

TABLE 3.10 **Flood damage in lakhs of rupees during 1953 to 1981 in the State of Uttar Pradesh**

Sl. No.	Year	Crops value (1)	Houses value (2)	Public utilities (3)	Total damage (cols 1+2+3)
1	2	3	4	5	6
1.	1953	1082.0	110.0	100.0	1293.0
2.	1954	682.0	19.0	150.0	851.0
3.	1955	2638.0	232.0	60.0	2930.0
4.	1956	2148.0	660.0	50.0	2858.0
5.	1957	1067.0	462.0	NA	1529.0
6.	1958	NA	NA	NA	NA
7.	1959	NA	NA	1.4	1.4
8.	1960	NA	NA	NA	NA
9.	1961	NA	2.7	27.0	29.7
10.	1962	1935.1	117.2	10.3	2062.6
11.	1963	1830.6	274.1	45.6	2150.3
12.	1964	1120.8	32.0	0.4	1153.2
13.	1965	11.0	1.0	1.0	13.0
14.	1966	690.90	70.8	NA	761.7
15.	1967	5555.1	500.3	105.3	6107.7
16.	1968	831.0	39.3	12.8	883.3
17.	1969	3460.1	276.5	108.7	3845.3
18.	1970	4300.0	2080.0	528.0	6903.0
19.	1971	10812.0	2025.7	4610.0	1747.7
20.	1972	284.8	101.0	NA	385.2
21.	1973	6745.4	1076.8	349.0	3171.2
22.	1974	5751.7	1278.5	3500.0	10530.2
23.	1975	3078.0	290.8	508.0	3876.8
24.	1976	18600.0	2250.0	3868.0	24718.0
25.	1977	7500.0	500.0	1660.0	9660.0
26.	1978	38200.0	9000.0	8936.0	56136.0
27.	1979	5176.9	172.2	NR	5349.1

TABLE 3.10 (*Contd.*)

1	2	3	4	5	6
28.	1980	19011.8	10012.4	7595.0	36619.2
29.	1981	36500.0	12277.6	11720.0	60497.6
	Total	179011.6	43862.2	43946.5	245027.6
	Maximum	38200.0	2277.6	11720.0	60497.6
		(1978)	(1981)	(1981)	(1981)
	Average	7160.5	1687.0	1910.7	9075.1

NR=Not Reported, NA=Not Available, Neg.=Negligible.

From : Central Water Commission, New Delhi—Personal communication.)

where many industries are located and establishment of new industries is rapid, especially in Kanpur, the industrial pollution problem is expected to be acute. In this region, the two urban agglomerations especially Kanpur and Allahabad are the ones where attention is needed, as their growth and industrial development has been rapid.

Kanpur is situated on the bank of the river Ganga which supplies drinking water to the city. Kanpur is considered to be a unique city. It is the largest city of U.P. and is one of the most important industrial towns of northern India with a strong commercial and industrial base. There are about a dozen textile mills, 60 tanneries, 3 woollen mills, 2 jute mills, Indian Explosives Limited, 2 thermal power stations and a number of small chemical, metallurgical and other factories. The effluents from these industries as well as the domestic effluents almost in untreated form, enter the stream of river Ganga at several points during its course. The problems of suspended and dissolved solids, oxygen demand and of chromium are expected to be severe in the region, though hardly enough data is available from the area. The course of the river has been fast drifting towards Unnao for the past several years. The narrow stream of water available near the bank of the river on the city side gets heavily polluted, especially during the period of October through June, when the flow of water is minimal. Most of the bathing Ghats situated along the bank of this stream are visited by millions of pilgrims regularly. The number of visitors swells tremendously on festive occasions like the Kumbh Mela. The analysis of effluents from different factories in Kanpur compared to the permissible standards for the disposal of the industrial effluents, speak of the high rate of pollution of the river at Kanpur [From Annexure V, Planning Commission (Education Division) U.O. No. Q—17022, 6/81=Edn. 1981]. Saxena *et al.*, (1966) reported some physico-chemical characters of the river water at Kanpur. The pH of the water was above 8.0, except during the rains. The highest pH value of 9.4 was reported due to textile wastes dumped in the river at the rate of 0.5 million gallons/day. The lowest pH value of 7.3 was obtained when sewage was dumped. Total solids ranged from 322-1090 mg/l and the turbidity ranged between 30 and 543 mg/l, which were mainly contributed by tanneries. Highest values were obtained in rains because of large quantities of silts being washed into the river. Alkalinity and hardness also increased with pH, though lower values were obtained in summer. Coliforms were found to be high in summer about 8 million MPN/100 ml. This reflected the extent of pollution caused by different sources along the banks of the river in this region.

Allahabad, the other urban agglomeration in the region is comparatively less industrialized as also less urbanized. Even then the city has not escaped from the pollution problem especially degenerating water quality of the two great rivers—Ganga and Yamuna along the banks of which the city is situated. Though the pollution load gets distributed along the two rivers, the problem of pollution does not disappear from the region. Studies by Chandra *et al.*, (1981) regarding pollution of Ganga due to the release of effluents from the IFFCO Complex, Phulpur, Allahabad district (25 km from Allahabad city), does reflect the acuteness of the problem. The study was undertaken on the pre-pollutional hydrobiological status of a stretch of river Ganga, likely to receive effluents from Indian Farmers' Fertilizer Cooperatives Ltd., (IFFCO) manufacturing urea (1550 tpd) and ammonium sulphate and to assess the effects of effluents on the ecological spectrum of the river in time and space. During the course of various manufacturing processes the fertilizer complex along with local township, consumes about 13,500 m^3 of water per day. About 6,500 m^3/day of

TABLE 3.11 **Characteristics of the composite effluent drained from major sections of IFFCO Complex on the bank of river Ganga near Allahabad**

Characteristics	Concentrations
Temperature (water)°C	30.5—32.0
Colour	Slightly brownish
Odour	Ammonical and Phenolic
Bicarbonate alkalinity as $CaCO_3$ (mg/l)	300 — 580.0
Carbonate alkalinity as $CaCO_3$ (mg/l)	Nil
Hydroxide alkalinity as $CaCO_3$ (mg/l)	Nil
Total alkalinity as $CaCO_3$ (mg/l)	300 — 580.0
Free ammonia (mg/l)	100.60 — 112.20
Nitrate-nitrogen (mg/l)	54.0 — 72.0
Chloride (mg/l)	15.80 — 29.60
Hardness as $CaCO_3$ (mg/l)	100 — 110.0
Sulphate SO_4 (mg/l)	800 —1000.0
Dissolved oxygen	Nil
Specific conductance (micromhos/cm)	1218 —1862.0
Suspended solid (mg/l)	340 — 560.0
Dissolved solid (mg/l)	1320 —1340.0
pH	8.6 — 8.9
Calcium as Ca^{++}	48 — 52.0
Turbidity (mg/l)	390 — 650
Oil (mg/l)	12.5 — 21.5

From : Chandra *et al.*, 1981

TABLE 3.12 **Number of organisms per litre of water in pre-pollution and pollution zones due to the release of composite effluents from IFFCO Complex on the bank of the river Ganga near Allahabad**

Sampling station	Pre-pollution zone				Pollution zone			
	I	II	III	IV	I	II	III	IV
Phytoplankton								
1. *Diatoms*								
Nitzschia	80	71	41	112	—	—	32	86
Synedra	302	166	209	620	—	—	—	110
Navicula	62	48	96	336	—	—	48	162
Gyrosigma	—	112	19	32	—	—	24	14
Pinnularia	—	27	13	132	—	—	92	27
Diatoma	23	48	—	16	—	—	18	—
2. *Blue green algae*								
Spirulina	37	96	60	226	34	68	153	109
Merismopedia	70	86	194	496	20	16	140	118
Anabaena	—	—	6	—	26	36	94	100
Microcystis	23	17	73	196	43	56	92	105
3. *Green algae*								
Zygnema	30	81	73	36	—	—	51	76
Pediastrum	92	—	160	288	—	—	47	150
Spirogyra	66	120	85	230	—	—	32	173
Ulothrix	184	176	496	624	—	—	14	592

Eudorina	116	140	118	80	—	—	64	56
Scenedesmus	146	121	84	88	—	—	33	22
Total	1231	1309	1727	3512	123	176	934	1900
Zooplankton								
1. *Rotifers*								
Brachionus	76	110	218	424	—	—	71	310
Keratella	68	81	122	210	—	—	90	174
Filinia	12	22	16	40	—	—	38	68
2. *Copepods*								
Cyclops	70	28	43	102	—	10*	92	140
Diaptomus	26	39	54	43	—	6*	72	110
Nauplius	18	23	26	70	—	—	8	67
3. *Cladoceran*								
Daphnia	—	10	—	29	—	2*	10	14
Total	270	313	479	938	—	—	381	883
Total Plankters	1501	1622	2206	4550	123	173	1315	2783

*=Dead zooplankton, Station I represents outfall, II—at 100 m, III—at 300 m and IV—at 500 m from outfall in the pollution zone (From Chandra *et al.*, 1981)

TABLE 3.13 **Quality of Ganga water during summer 1977 in the region of the Upper Ganga Plain (u/s=upstream, d/s=downstream of the towns)**

Point	Temperature °C	DLP cm	DO mg/l	BOD mg/l	Conductivity mmhos/cm	Chlorides mg/l	Hardness as $CaCO_3$ mg/l	Coliform MPN no/100 ml
Narora u/s	30	24	7.5	4.2	180	4	84	130
d/s	29	28	7.4	7.2	147	5	76	63
Kannauj u/s	28	39	6.0	4.6	301	7	140	240
d/s	29	22	5.9	7.0	315	7	136	130
Kanpur u/s	27	19	7.5	10.0	287	8	108	35
d/s	29	18	6.0	27.0	293	9	120	2400
Allahabad u/s	31	31	6.5	6.0	406	16	134	350
d/s	30	35	6.5	6.8	406	24	138	2400

From : Bhargava, 1981a.

TABLE 3.14 **Water Quality Indices (WQI) during summer and classification of Ganga waters for different uses in the Upper Ganga Plain (u/s=upstream, d/s=downstream of the towns)**

Point	Beneficial uses										
	Drinking without treatment, bathing		Public water supplies		Agriculture		Industrial		Fish culture		Overall WQI
Narora u/s	55	III	68	II	94	I	57	III	72	II	69
d/s	52	III	64	III	94	I	61	III	66	II	67
Kannauj u/s	56	III	69	II	89	II	61	III	71	II	69
d/s	45	III	58	III	88	II	51	III	63	III	61
Kanpur u/s	41	III	51	III	89	II	50	III	65	II	59
d/s	11	IV	20	IV	86	II	49	III	23	IV	38
Allahabad u/s	50	III	62	III	82	II	56	III	66	II	63
d/s	35	III	49	III	78	II	57	III	66	II	57

I=>90—Excellent, II=65—89—Good, III=35—64—Satisfactory, IV=11—34—Poor,

V=<10—Unacceptable.

From : Bhargava, 1981a.

waste water is discharged as effluent from the fertilizer complex. The composite effluent flows through a 26 km long stream into river Ganga. On 10th October, 1980, fertilizer wastes were discharged for the first time into the river Ganga through Mogarsan tank and Bairagiya nallah for a short spell of ten days. The composite effluent was characterized by high free ammonia content, alkalinity, nitrate nitrogen, etc. (Table 3.11). Heavy fish mortality (50 quintals) of major and minor carps, catfishes and murrels weighing 200 gm to a kg each, and damage to two farm animals in a stretch of 16 km from the point of discharge were reported. Fish mortality could be attributed to discharge of free ammonia at fairly alkaline pH, while damage to livestock was ascribed to the high nitrate nitrogen content of the effluents. The general effect of toxic discharge was witnessed up to 300 m in the river. Plankton population was seen to disappear from the outfall region in contrast to its continued concentration up to 4550 No /l recorded prior to release of effluents (Table 3.12). The benthic community was also conspicuous by its total absence for 300 m. It would prove educative to examine the different treatment methods for rendering the wastes innocuous before their release into the Ganga river. Bhargava (1981) using a summer data on river Ganga extending from Rishikesh to Varanasi has prepared water quality indices (WQI) for different spots along the Ganga river. The data about the quality of river water and water quality indices are given in Tables 3.13 and 3.14. He further classified the Ganga for various individual uses and also for the overall use, as given in Table 3.14 which was based on the following ranges of WQI.

WQI	Range of WQI		Classification of each use
100	Greater than 90	I	Excellent
80	Between 65 & 90	II	Good
50	Between 35 & 64	III	Satisfactory
20	Between 11 & 34	IV	Poor
1	Below 10	V	Unacceptable

It is clearly infered that the water quality level needs continuous examination. The effluent wastes from industries also need treatment before they are discharged into the river.

CHAPTER IV

The Middle Ganga Plain

4.1. LOCATION

THOUGH not a clear-cut physical unit, the Middle Ganga Plain (24° 30′ N—27° 50′ N and 81° 47′ E—87° 50′ E) is a large physical area (144,409 km^2). It covers Eastern Uttar Pradesh (mostly Purbia Plain) and the Bihar Plains in their entirety, lying on either side of the Ganga and the Saryu (Ghaghara) within the Himalayan and the Peninsular ramparts on the north and south respectively (Fig. 4.1). It is a transitional region par excellence, interposed in the enormity of the Ganga Valley, and all its physical, human, cultural and economic fabric seems to have been woven out of the various unique phenomena interplaying in transition.

The northern limit is provided by the Indo-Nepal International boundary which broadly cuts through the Bhabar and Tarai zone down the Siwaliks; only in the northern part of Champaran (Bihar) a small chunk of the Siwaliks is included in this region. In the south, the 150 m contour marks the boundary and skirts down the bold face of the hill-spur-scrap and alluvial embayment zone of the Vindhyan-Rohtasgarh-Chhotanagpur Plateaux, thus excluding most of the Mirzapur district, Chakia tahsil of Varanasi district, Kaimur-Rohtasgarh plateau and, further east. only the southernmost strips of the Gaya, Monghyr and Bhagalpur districts and all the sub-divisions of Santhal Parganas except Godda. The eastern boundary of the Upper Ganga Plain marks the western limit of this region, while the eastern limit corresponds to the Bihar-Bengal State boundary except the Kishanganj sub-division in the NE of Purnea which has been included in the Lower Ganga Plain (Singh, 1965 a).

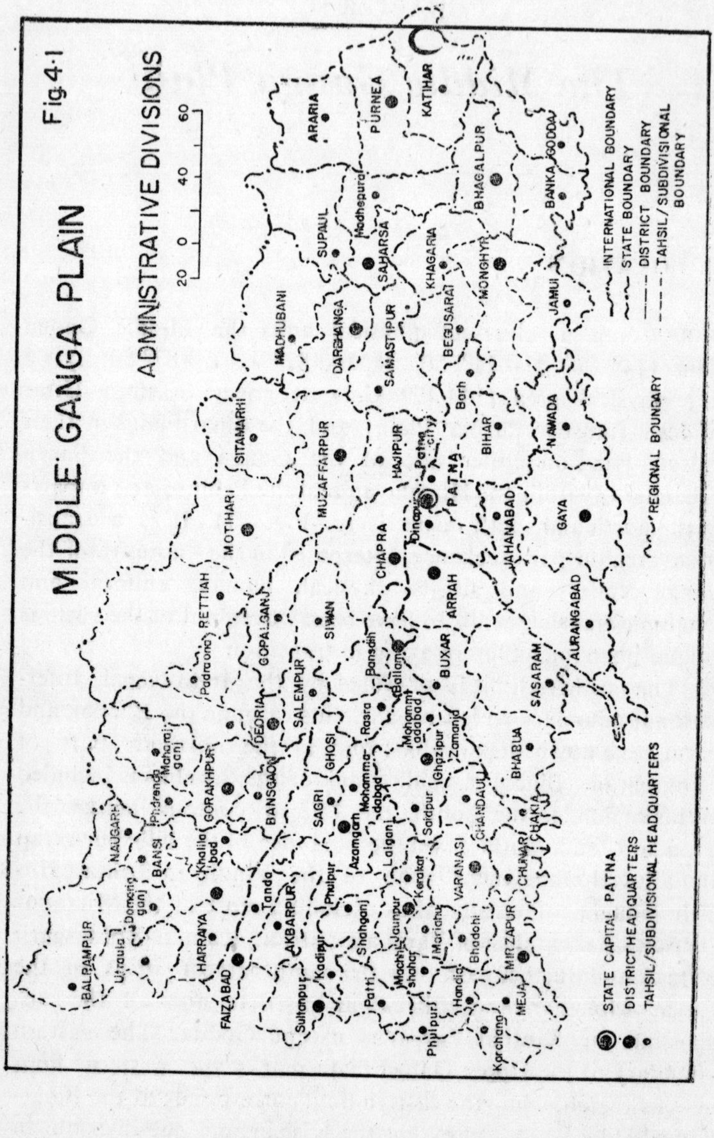
MIDDLE GANGA PLAIN
Fig.4·1
ADMINISTRATIVE DIVISIONS
20 0 20 40 60
INTERNATIONAL BOUNDARY
STATE BOUNDARY
DISTRICT BOUNDARY
TAHSIL/SUBDIVISIONAL BOUNDARY
REGIONAL BOUNDARY
STATE CAPITAL PATNA
DISTRICT HEADQUARTERS
TAHSIL/SUBDIVISIONAL HEADQUARTERS
BALRAMPUR
FAIZABAD
AKBARPUR
GORAKHPUR
DEORIA
GOPALGANJ
RETTIAH
MOTIHARI
SITAMARHI
MADHUBANI
SUPAUL
ARARIA
PURNEA
KATIHAR
BHAGALPUR
BANKA
GODDA
SAHARSA
KHAGARIA
MONGHYR
JAMUI
BEGUSARAI
SAMASTIPUR
DARBHANGA
MUZAFFARPUR
HAJIPUR
PATNA
BIHAR
NAWADA
GAYA
JAHANABAD
AURANGABAD
SASARAM
BHABUA
ARRAH
BUXAR
CHAPRA
SIWAN
SALEMPUR
GHOSI
SAGRI
Azamgarh
VARANASI
CHANDAULI
CHUNAR
CHAKIA
MIRZAPUR
MEJA
Sultanpur

The Middle Ganga Plain includes the administrative divisions of Gorakhpur and Varanasi (excepting part of the Chakia tahsil of Varanasi district and major portion of Mirzapur district) and tahsils of Balrampur and Utraula (Gonda), Faizabad, Tanda and Akbarpur (Faizabad), Sultanpur and Kadipur (Sultanpur), Patti (Pratapgarh) and Phulpur, Handia, Karchhana and Meja (Allahabad) in Eastern Uttar Pradesh and divisions of Tirhut, Bhagalpur (excluding Kishanganj sub-division and including only the Gonda sub-division) and Patna, leaving out the area above 150 m in the south. The maximum length from east to west is about 600 km while its width from north to south approximates 300 km.

4.2. THE PHYSICAL SETTING

4.2.1. Physiography

Structurally the region is a segment of the great Indo-Ganga trough; and some marginal portions of the other two major formations, i.e., the Siwaliks in the northern part of Champaran district and the fringes and projections of the Peninsular Block in the south. The Himalayan foot-hills region of the Siwaliks formation covers an area of about 364 km^2 and continues across the Indo-Nepal boundary towards north. Between the relatively lower and shorter Ramnagar Dun Hill Range and the northern, higher and longer Sumeshwar Range lies the Harsha Valley, called the Dun, about 22 km long and less than 150 m in elevation.

The average alluvial filling varies widely from one part to another; particularly the depth increases up to 8-10,000 m as the Himalayas are approached, while it thins out as a mere veneer on the Peninsular margins. It is indicated that the entire region has suffered great down-warping due to the Himalayan upheaval.

The alluvial sediments have less 'kankar' formation because of more riverain character of the plain, and more particularly of the higher incidence of 'khadar' lands.

The region is almost of a homogeneous level, seemingly featureless plain from one end to the other. In general, it is below 100 m above the sea level. In the east the Kosi Plain ranges between 30 m in the south and 75 m in the extreme north.

4.2.2. Drainage

Drainage lines hold particular significance in the region: not only do they provide redeeming topographic breaks in the general flatness of the plain and provide sub-regional or even local uniqueness and individuality to the different parcels of land, but they also govern to a great extent, the human occupance of land, particularly the agricultural land and settlements.

The drainage pattern is generally dendritic and the common characteristic feature available throughout the plains is that the rivers meet at acute angles, and several tributaries form parallel or sub-parallel lines to the main streams (Fig. 4.2).

The Ganga is the master stream and is the recipient of all water-lines in the region. The regional drainage can be studied through the subsystems: (i) the Ghaghara system, (ii) the Gandak system, (iii) the Kosi system, all north of Ganga, (iv) the Son system, and (v) the other tributary drainage to the Ganga from the south.

The Ganga, a snow-fed Himalayan river and heavily leaden with detritus, flows sluggishly from west to east, through the central-southern part of the region, being closer to the Southern Uplands than to the Himalayan wall. While its northern perennial tributaries have a wider arena to carve out their flat valleys through their highly meandering courses, the southern tributaries meet the Ganga rather hurriedly because of steeper slope, particularly east of Monghyr and west of Roh'as. The average gradient is 9.5 cm per km in the region, slowing down to 6 cm in Bihar. The flood-plain varies from 5 to 30 cm or more and is known as 'khadar' including sometimes vast 'diara' lands between the occasionally braided channels and also sand flats, all subject to annual inundation. The Ganga banks are infested with stable and high (10 to 15 m) levees intercalated with 'kankar' gravel or other resistant rock reefs as at the sites Mirzapur, Varanasi and Patna. Northernmost projections of hard rock basements of the southern uplands is another type of bank feature, e.g., sand-

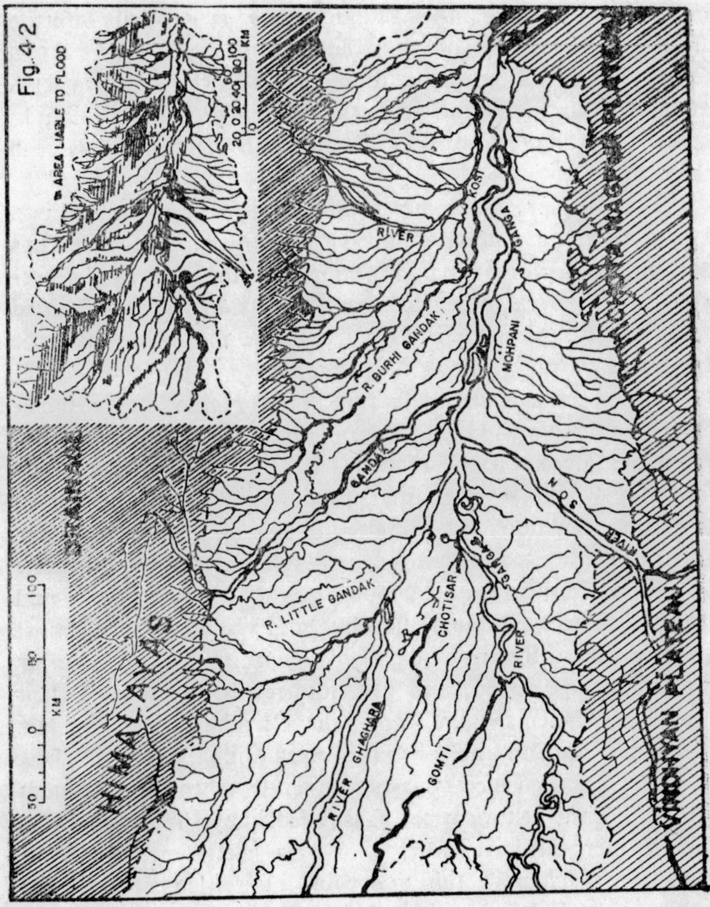
Fig. 4.2
AREA LIABLE TO FLOOD
KM
HIMALAYAS
R. LITTLE GANDAK
GANDAK
R. BURHI GANDAK
RIVER
KOSI
RIVER GHAGHARA
CHOTISAR
GOMTI
GANGA
RIVER
SON
MOHPANI
GANGA
VINDHYAN PLATEAU
CHOTA NAGPUR PLATEAU

stone at Chunar and quartzites at Monghyr. Sometimes the projections appear in the mid-stream, as at Sultanganj. As such, the southern bank is relatively more permanent and stable than the northern one.

The Kosi is the 'sorrow' of Bihar and the wildest and the most devastative of the Indian rivers. It deposits infertile sediments consisting of micaceous sands and renders vast fertile lands into sandy and marshy flats. The Kosi project is an attempt to tame and train the river. A protective embankment from Bagha to its confluence with the Ganga has been thrown.

The Ganga receives numerous tributaries from the southern Uplands among which the Son is the largest. The river has been notorious for its changing courses in the past, as is traceable from several old beds on its east, but has been tamed squarely with the anicut at Dehri, and now more so with the Indrapuri Barrage. a few km upstream.

Floods are a recurring feature in the region, more particularly in the North Ganga Plain (Table 4.3). Not infrequently, when the Ganga itself remains in spate, and thwarts the incoming water from its numerous tributaries in the region, large areas get flooded leading to breaks of transport system and devastation of life and property. Defective and inadequate provision of culverts and syphons in the transport nets and canals add to the frequent flooding and waterlogging. Almost all the rivers in the region develop a capacity to spill over in the monsoon period, and are notoriously dynamic in their character, particularly the Rapti, the Ghaghara, the Gandak, the Kosi, the Son and the Ganga itself (Singh, 1965 b). These rivers have shifted their courses frequently, covering wide areas throughout the historic times as is reflected by the remnants of their former beds in the form of ox-bow lakes, meander loop, dead arms, 'chaurs', tals and ruins of settlements, etc. For instance, the Ganga has shifted over 35 km between Bhojpur *tal* and Surha *tal* causing changes in the site of Ballia. Towns and river ports like Gola, Barhaj, etc., have been affected by similar shifts of the Ghaghara. The Kosi, the most dynamic among the rivers, has marched about 120 km during the last two hundred years from near the town of Purnea up to the Tiljuga within Darbhanga district. Historically, it has oscillated from the Brahmaputra to about the Ganga-Gandak confluence.

4.2.3. Climate

The lie of the land between the Himalayas on the north and the Peninsular foreland in the south, and lack of physical undulations to check the sweeping winds and air-currents from the east and west, conspire to make the region only transitional in character between the relatively drier Upper and perhumid Lower Ganga Plains (vide Tables 4.1 to 4.3). The winter cyclones with their occasional cold waves persist sometimes for a week or more, as in 1961. Hot summer winds from the west sweep the entire region almost unhindered. The duststorms are rainier here than in the Upper Plain but much less than in the Lower Plain, as also the winters are less rainy than in the Western Plains but more than in the Eastern counterpart.

Although the region has somewhat continental interior location within the subtropical climatic belt, the monsoon reigns supreme here, and carries great weight in the overall human occupance pattern and economic development.

The season of general rains starts abruptly with cloud burst of the summer (SW) monsoon, heralding the active monsoon period usually by the 7th of June in the N and NE Bihar, by 15th June in W. Bihar and the third week of June in the Eastern U.P. Plain. The advent of the monsoon is occasional by almost a sudden rise of relative humidity to over 70%. a sudden fall in temperature, and sudden change of wind direction from the westerly and north-westerly to easterly and south-easterly. The south west monsoon gradually retreats from the region by the end of September or first week of October. October, however, remains a sticky and more oppressive sultry month under a comparatively clear sky, the relative humidity still marking high and the moderating rains almost gone.

4.2.4. Soils

Apart from the undifferentiated soil of the Siwalik fringe zone in Champaran district, most of the region has broad

TABLE 4.1 **Average seasonal rainfall (Percentage of annual total and annual in cm) in parts of the States situated in the Middle Ganga Plain**

States	Winter (Dec-Feb)	Summer (March-May)	Monsoon (June-Sept)	Post-Monsoon (Oct-Nov)	Annual (in cm)
Bihar (North)	2.9	6.8	85.0	5.3	122.6
Eastern U.P.	3.9	2.9	88.0	5.2	99.3
Western U.P.	6.0	3.6	87.8	2.6	95.6

(Table based on Ramdas, L.A, "Weather and Crops", Handbook of Agriculture, Indian Council of Agricultural Research, New Delhi, 1961—from Singh, 1971)

TABLE 4.2 **Normal mean daily maximum and minimum temperature (°C) in parts of the States situated in the Middle Ganga Plain**

Months	Normal mean daily maximum temperature (°C)			Normal mean daily minimum temperature (°C)		
	Bihar (North)	U.P. East	U.P. West	Bihar (North)	U.P. East	U.P. West
Jan.	23.3	23.3	21.7	10.0	8.9	7.8
Feb.	26.1	26.1	34.1	12.2	11.1	10.0
Mar.	32.2	31.7	30.0	17.2	15.6	14.4
Apr.	37.2	37.8	36.7	22.2	21.7	20.6
May	37.8	40.0	40.0	25.8	25.6	25.0
June	35.0	37.0	37.8	26.1	27.2	26.7
July	32.8	32.8	32.8	26.1	26.7	26.1
August	31.7	32.2	32.2	25.6	26.1	25.6
Sept.	31.0	32.8	32.8	25.6	25.0	23.3
Oct.	31.7	32.8	32.8	22.2	20.0	17.8
Nov.	27.8	28.3	27.7	15.0	12.8	11.1
Dec.	23.9	23.3	23.3	10.0	8.9	7.8

From : Singh, 1971.

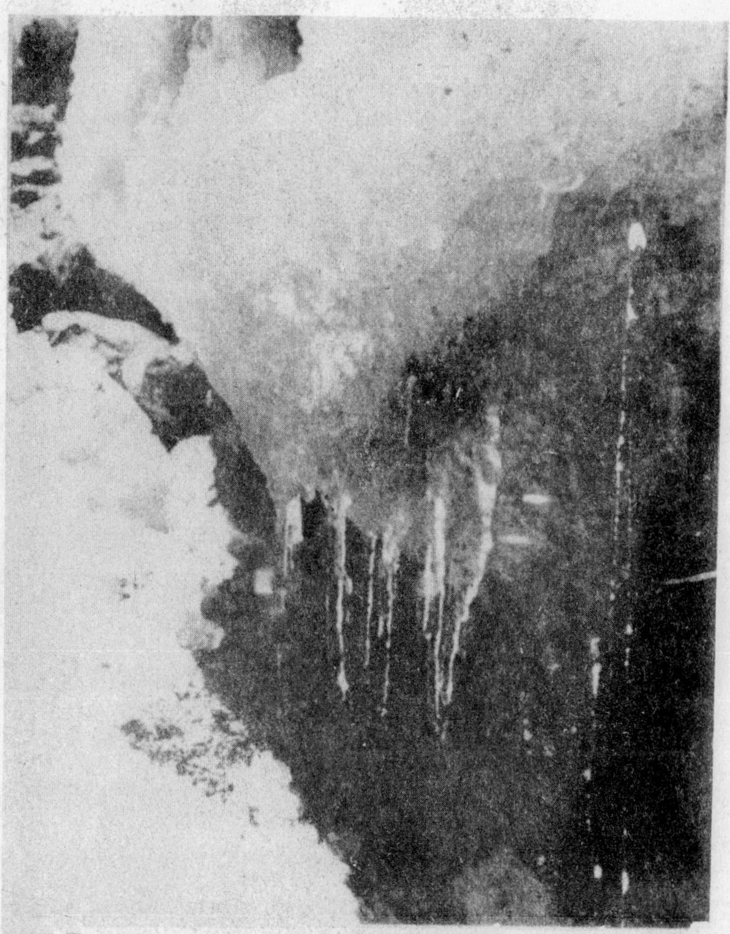

1. Gomukh Glacier : Source of River Ganga.

2. Gangotri, a temple dedicated to Ganga and a great pilgrimage, about 3 miles from the source of the river in the Gomukh Glacier.

3. Unmindful of what he is doing this lumberjack axes a pine tree in Himachal Pradesh. Such illegal felling continues unabated.

4. Banaras (people taking bath)

5. Destruction of Chir pines in Western Himalays.

6. Debris from a quarry at Mallah (Haryana), being disposed off into the Ghaggar river.

7. In the early morning mist and cold lakhs of pilgrims assembled for a holy bath at Kumbha Mela, Allahabad.

8. Kumbha Mela, Allahabad.

10. View of a Ghat at Banaras.

11. A view of the Ghat at Banaras.

13. Banaras : Ghat.

14. Bhaghirathi floods in Uttarkashi in U. P.

15. Villages marooned in the floods in U. P.

16. The Gomati in fury near Hanuman Bridge in Lucknow. (PANA Photograph).

17. Towing away the belongings in flooded Patna.

18. Floods in Patna (Indira Gandhi watching the flooded area from a helicopter).

19. Homes were submerged in Midnapore village.

20. Flood—affected people evacuated from villages in Danapur.

21. 60 miles northwest of Calcutta, Nabdwip—95,000 people made home-less in the area and business brought to a standstill.

RSA—1. Men and buffaloes having respite in water on a summer noon.

RSA—2. Extensive use of Ganga Ghats by washermen adds detergents and alkalies.

RSA—3. Vultures on a floating human cadaver.

RSA—4. Vultures on animal carcass, floating in the Ganga river.

TABLE 4.3 **Rainfall in cm, mean deviation and number of abnormalities in parts of the States situated in the Middle Ganga Plain**

States	Rainfall in cm		Number of abnormalities			Limit for defining abnormality
	Normal (1940)	Mean deviation	Floods	Droughts	Total	
Bihar North	102.5	16.25	7	8	15	10
U.P. East	87.5	16.00	10	7	17	10
U.P. West	82.5	15.75	4	8	12	10

(The years in which the actual rainfall exceeds twice the mean deviation are the flood years and those in which the actual rainfall was less than the normal by more than twice the mean deviation are the droughts years.) (From Singh, 1971).

alluvial soil cover (Fig. 4.3). The soils are rich in mineral and organic plant foods. There is, however, a serious deficiency in nitrogen, which may be improved upon by the use of compost, manures and fertilizers.

4.2.5. Natural vegetation

An almost unhindered human occupance for over three millenia and centuries of plough and pastoral culture has affected the natural vegetation in the region except in the northern part of Champaran and pockets of the Tarai and some river banks. With a moderate rainfall and fertile soil the region is a natural habitat of a dense forest cover of Sal and other species like Shisham, Jamun, Mahua, Ber, etc. Even up to the 1840's a large area of the Saryupar was densely forested, and so also were the banks of the Rapti, the Ghaghara and other rivers.

The remnants of the once extensive sal forests are found in northern Gorakhpur, Saharsa and Purnea districts. Elsewhere, the induced vegetation is seen in the form of a savannah

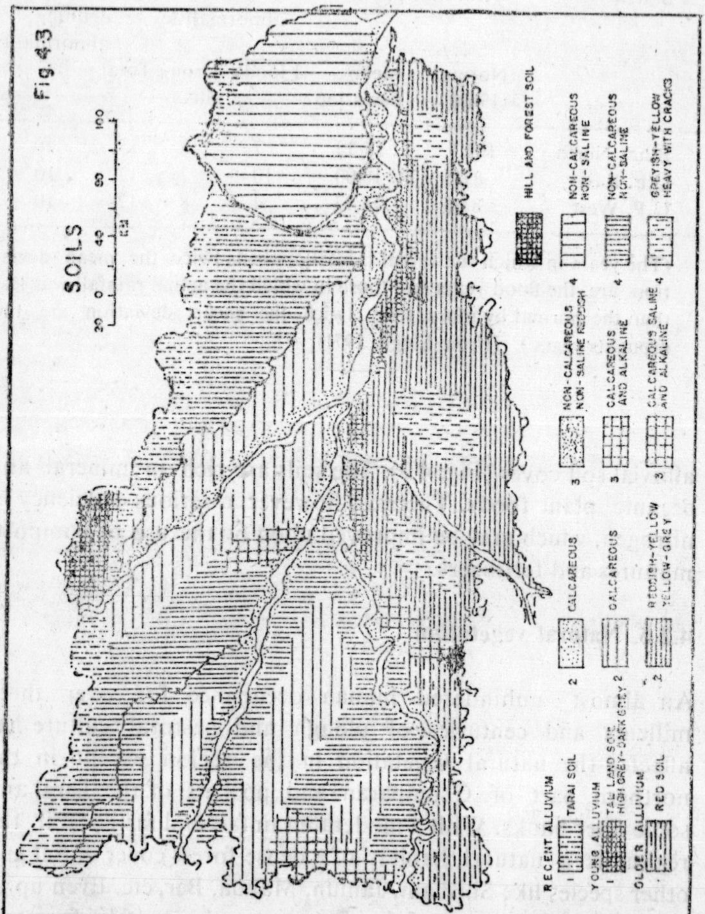
Fig 4.3
SOILS
20 40 60 80 100
KM
HILL AND FOREST SOIL
RECENT ALLUVIUM
1 TARAI SOIL
2 CALCAREOUS
3 NON-CALCAREOUS NON-SALINE REDDISH
4 NON-CALCAREOUS NON-SALINE
YOUNG ALLUVIUM
1 TAL LAND SOIL HIGH GREY-DARK GREY
2 CALCAREOUS
3 CALCAREOUS AND ALKALINE
4 NON-CALCAREOUS NON-SALINE
OLDER ALLUVIUM
1 RED SOIL
2 REDDISH-YELLOW YELLOW-GREY
3 CALCAREOUS SALINE AND ALKALINE
4 GREYISH-YELLOW HEAVY WITH CRACKS

with grasses and bushland dotted with trees of different sizes. The village westlands or vacant places are covered by naturally growing trees like the peepal, banyan, tamarind, mahua, neem, babul, palmyra, date-palm, ahak, etc., while such grasses as bher, moonj, kans, jhau, etc., are found in the 'diaras', 'Tarai' or other low-lying areas. The more omnipotent dub, motha, and other less pervasive types are found in the cultivated fields and mounds or field-boundaries and left-over lands. The planted fruit orchards and groves of trees consist of mango, jamun, guava, mahua, jackfruit, plums, lemon, etc., and also timber trees like shisham etc., while bamboo clumps grow around and within settlements. Also such thorny-bushy plants as the senhur, plum bushes, etc., grow on waste or neglected lands, the former is sometimes planted for fencing farms or orchards.

The Champaran forests are classified as (1) Siwalik sal, (2) Moist (Gangetic) high alluvial sal, and (3) Gangetic moist deciduous riverain forest.

There is some difference in vegetal cover north of the Ghaghara and the Ganga on the one hand and to their south on the other. The groves and orchards, marshy vegetation, etc., are more frequent, particularly in N. Bihar Plains. In the south, there is more open parkland vegetation and the frequency of groves decreases. Isolated trees are more common here. 'Sabai' is the most important grass of the region, particularly in the north.

In most of the districts like Ballia, Ghazipur, Jaunpur, Azamgarh in the lower Ganga-Ghaghara Doab and Saran, Muzaffarpur, Darbhanga and Saharsa in N. Bihar Plain, forest cover is particularly non-existent (below 0.01% of the total area of the district), while in most of the others it is negligible (below 1%) as in Purnea, Pratapgarh and Faizabad, Deoria, Sultanpur, Patna and Basti. Some parts of the more humid and riverain northern tahsils of Saryupar and other detached pockets of Gorakhpur, Gonda and Bahraich districts, have appreciable forest cover and swell the district figures in per cent of the total forest area to 8.7, 9.1 and 14.3 respectively. Champaran also has a significant forest cover (10.92%). In the districts south of the Ganga, the forests occur mostly outside the plains (From Singh, 1971).

4.2.6. Ground water

The Middle Ganga Plain with adequate rainfall and generally porous alluvial soil cover seems to have adequate storage of ground water, but needs detailed scientific survey to locate and measure their incidence and tapability. This reservoir occurs in the aquifers and water-bearing sediments. The water table is generally lower in the newer alluvium, particularly in the river levee zones where the sandy nature of the sub-soil allows greater sinking of the water than in the clayey interfluvial areas of older alluvium. The water table varies from 3 to 20 m, being particularly high in Tarai zone. There also exist somewhat artesian conditions, probably at a depth of 60-65 m below the surface, and actually some artesian wells have been sunk in the U.P. Tarai.

4.2.7. Physical resource base

The region derives its resources primarily from the rich fertile soils and surface and ground water resources of the vast plains. It also has vast human resources enriched by a cultural heritage of millenia of years.

The region like its eastern and western counterparts, is endowed with a vast, well drained and almost isotropic land with one of the richest soil covers in the world. With its all-the-year-round growing season and adequate annual rainfall, it can grow a variety of food and industrial crops with the aid of supplemental ground and surface water resources that can be stored and channelised in the rain-deficient months or seasons. The region suffers from droughts and floods almost alternately and the twin ills, in addition to soil erosion, can be easily remedied by the introduction of a well-articulated irrigation-cum-drainage system and reafforestation and land management schemes. The region has, from times immemorial, developed an indigenous system of tapping the upper surface water resources through wells, and at present hundreds of tubewells are constructed to utilize the vast, deeper channels of the rich ground water reservoir. The forest cover is inadequate, which has not only led to appallingly short supply of fuel wood and timber but has also accelerated the process of soil erosion and drainage disability. Only some pockets in northern Tarai and small patches elsewhere are natural vegetal cover. By destroying the local forests and woodlands on the village commons or wastelands and inter-village boundary lands, the plains people, for the most part, have deprived themselves of their local fuel wood and timber supply for their domestic and agricultural needs. Bamboo, and such industrial trees as mango, mahua, jack fruit, plum and shisham are available only in private possessions, rendering some service for timber. In places, long grasses and cane-like reeds grow which are utilized for making furniture, containers and similar equipment for local use and as thatching material.

The mineral wealth is negligible. Clay and sands are used for making walls through dried cakes or baked-bricks for pottery and tiles while the 'kankar' is utilized for roads as well as for manufacturing lime.

4.3. THE CULTURAL SETTING

4.3.1. Population

With a total population load of 83.17 million (74.8 million rural and 8.34 million urban) on an area of 154 thousand km^2, the Middle Ganga Plain is conspicuous on the population map of India, due to very high overall density of about 593 persons per km^2. This includes little urbanism (10.03%) (Table 4.4). Rarely in the world except perhaps in China or Bangla Desh such a high population density with a rural base is to be so evenly met with, over such an extensive region.

The overall density of population is evenly high excepting in the northern Siwalik chunk, southern plateau fringes and a few pockets of the forested and ill-drained Tarai, marshy, riverine flood-prone patches, 'khadar' stripe of larger rivers and 'Usar' infested patches of the lower Ganga-Ghaghara Doab, all of which have comparatively lower density. The high density reflects upon the overall isotropic nature of the plain, suitability of homogeneously diffused habitable conditions, fertility of soil, drainage conditions, tapability of potable ground water resource through wells and borings, high

TABLE 4.4 **Distribution of rural-urban population for each district situated in the Middle Ganga Plain**

District	Area 000's km²	Population 1981		Density per km²	Percentage of urban to total population		Percentage growth rate, 1971-81	
		Total	Urban		1971	1981	Total	Urban
1	2	3	4	5	6	7	8	9
Uttar Pradesh								
Faizabad	4.5	2,369,626	259,799	525	9.56	10.96	+22.95	+41.06
Sultanpur	4.4	2,057,783	67,379	459	1.97	3.31	+24.03	+108.41
Basti	7.2	3,576,783	171,867	495	2.52	4.81	+19.86	+128.25
Gorakhpur	6.3	3,795,735	400,836	605	7.90	10.56	+24.93	+66.91
Deoria	5.4	3,487,350	231,427	640	2.96	6.64	+24.00	+178.46
Azamgarh	5.7	3,532,876	317,542	615	5.21	8.99	+23.64	+113.31
Jaunpur	4.0	2,527,012	168,720	626	6.21	6.68	+26.01	+35.45
Ballia	3.2	1,926,267	176,080	604	4.58	9.14	+21.23	+142.00
Ghazipur	3.4	1,941,516	154,238	575	4.50	7.94	+26.76	+123.53
Varanasi	5.1	3,696,768	990,798	726	25.13	26.80	+29.60	+38.23
Mirzapur	11.3	2,033,834	267,497	180	12.03	13.15	+31.97	+44.29

Bihar								
Patna	3.2	3,031,463	1,117,737	951	30.22	36.87	+34.68	+64.30
Nalanda	2.4	1,638,364	223,092	692	9.30	13.62	+25.44	+83.61
Nawada	2.5	1,099,067	73,044	441	5.59	6.65	+22.95	+46.07
Gaya	6.6	3,138,316	339,719	479	9.41	10.82	+23.19	+41.67
Aurangabad	3.3	1,236,528	85,805	374	4.94	6.94	+21.69	+70.92
Rohtas	7.2	2,363,905	230,070	328	7.89	9.73	+21.61	+50.07
Bhojpur	4.1	2,389,658	257,636	583	8.55	10.78	+19.77	+51.04
Saran	2.6	2,074,267	169,428	783	5.76	8.17	+21.32	+72.18
Siwan	2.2	1,778,050	78,352	801	3.54	4.41	+21.61	+51.19
Gopalganj	2.0	1,361,555	67,788	670	2.65	4.98	+22.94	+131.29
Paschim Champaran	5.2	1,967,579	144,559	376	7.38	7.35	+23.98	+23.37
Purbi Champaran	4.0	2,427,749	112,853	612	3.45	4.65	+24.11	+67.09
Sitamarhi	2.6	1,929,827	87,282	730	3.16	4.52	+21.94	+74.77
Muzaffarpur	3.2	2,353,283	189,765	742	6.62	8.06	+23.27	+50.16
Vaishali	2.0	1,661,512	107,640	816	5.76	6.48	+23.17	+38.64
Begusarai	1.9	1,456,512	153,687	759	9.22	10.55	+26.94	+45.22

TABLE 4.4 (*Contd.*)

1	2	3	3	5	6	7	4	9	10
Samastipur	2.9	2,116,741		88,026	729	3.32	4.16	+23.13	+54.16
Darbhanga	2.3	2,004,583		175,879	880	8.14	8.77	+23.53	+33.18
Madhubani	3.5	2,324,074		72,307	664	2.26	3.11	+22.83	+68.80
Saharsa	5.9	2,952,234		168,929	500	4.53	5.72	+25.61	+58.66
Purnia	7.9	3,592,637		286,018	452	6.06	7.96	+28.09	+68.32
Katihar	3.1	1,427,197		134,286	467	7.05	9.41	+25.52	+67.60
Monghyr	7.9	3,314,806		466,285	419	12.98	14.07	+20.75	+30.89
Bhagalpur	5.6	2,610,719		303,389	467	10.61	11.62	+24.85	+36.74
Middle Ganga Plain	154.6	83,176,176		8,339,759	593	7.46	10.03	+24.22	+68.57

From : Appendix I of Paper 2 of 1981—Rural-Urban Distribution, and Appendix III of Paper 3 of 1981, Workers and Non-workers. Census of India, Series 1, 1981.

percentage of arable land and generally good conditions for making a living out of the local and regional resources.

The density per km^2 varies from 180 (Mirzapur district in the East U.P. Plain) to 951 (Patna in Bihar Plain), but the occurrence of such extreme density categories is limited. About half the region covering a solid block from the Kosi westwards, through the entire North Bihar Plain (except northern Champaran), eastern half of the Saryupar, major portion of the lower Ganga-Ghaghara Doab east of Mirzapur-Jaunpur-Azamgarh line, western and northern half of the south Bihar Plain together with Chandauli tahsil and part of Chakia in Varanasi district carry a density of over 600 persons per km^2. The hard core of this solid block lies around the Ganga-Gandak-Son and Ghaghara confluence point with a density of over 700 persons, as is reflected by the district densities of Patna, Saran, Muzaffarpur and Darbhanga.

There is a general decrease in density towards the west corresponding with decreasing rainfall. The erratic behaviour of the Kosi may be responsible for the low population density in the Kosi Plain in the east.

4.3.1.1. Tredd of growth

There has been a remarkable growth of population during the last couple of decades. The average growth rate for the region was found to be 24% for the decennial 1971-81. The highest growth rate of 35% was recorded for the period in Patna district, while the lowest was 19.77% in Bhojpur district (Table 4.4).

The region had registered an actual decline of absolute numbers in the decade 1911-21, due to epidemics in 1911-12, followed by two successive droughts, floods and famine. The troubles were all the more accelerated by famine, diseases and epidemics in 1918-19, specially influenza. The denser pockets often suffered the most. During 1921-31 there was a spurt of growth but the trend got accelerated since the thirties, making an average annual growth rate of about 1.4% consistently.

Reclamation of Tarai tracts in Gorakhpur, Champaran and other areas, arresting of floods in the Kosi Plain, and introduction of irrigation facilities have led to agricultural settlements, stabilization and economic prosperity, thus leading to constant population growth, particularly during the post-independence period. The high urbanization rate in some areas and a remarkable decline in death rate in general and better health facilities have also contributed to this trend. The highest and lowest increases in growth rates for 1971-81 decennial in the urban population, were recorded for the districts of Deoria (178.46%) and Paschim Champaran (23.37%) (Table 4.4).

4.3.1.2. Population structure

Agriculture is the mainstay of the people in the Middle Ganga Plain, engaging 81.67% (47.21% cultivators and 34.46% agricultural labourers) of the total working population. The remaining portion of the working population was made up of workers in household industry (3.81%) and in other services (14.69%) (Table 4.5).

Although the region is mainly agrarian in nature, there was a drop of about 2.58% in its agricultural workers in the decennial 1971-81. This shift though small in magnitude, showed the process of slow urbanization taking place in the region. This compared with the 2.57% increase in the urban population between 1971 and 1981.

4.3.1.3. Urban settlement

Although the region is one of the least urbanized parts in the country, it contains the oldest living city of the country, and perhaps of the world, i.e., Varanasi, which epitomizes the history of the human settlements in the Middle Ganga Valley. Some of the very ancient towns in the region are Ayodhya, Ballia, Gadhipur (Ghazipur), Vindhyachal, Chunar, Kushbhawanpur (Sultanpur) in East U.P. Plain and Pataliputra (Patna), Gaya, Champa (Bhagalpur), etc., in Bihar Plains. The Ganga Valley, its own banks and confluences with other rivers, have given rise to many of these ancient towns as the river has been thought of as sacred and sanctimonious, as a highway for commerce, politics, accessibility and also good for defence.

TABLE 4.5 **Distribution of main workers by cultivators, agricultural labourers, household industry and other workers in each district situated in the Middle Ganga Plain**

District	Percentage of workers							
	Cultivators		Agricultural labourers		Household industry		Other workers	
	1971	1981	1971	1981	1971	1981	1971	1981
1	2	3	4	5	6	7	8	9
Uttar Pradesh								
Faizabad	56.64	63.02	26.60	18.18	4.53	4.51	12.23	14.29
Sultanpur	56.47	63.28	31.65	23.44	3.32	3.01	8.56	10 27
Basti	64.77	71.34	26.00	16.40	2.52	3.18	6.71	9.08
Gorakhpur	46.65	52.98	37.85	26.10	2.10	3 16	13.40	17.76
Deoria	57.69	63.42	31.57	20.53	1.77	2.83	8.97	13.22
Azamgarh	54.42	61.26	27.95	17.92	6.96	8.02	10.67	12.80
Jaunpur	59.77	66.27	23.79	13.19	3.74	5.47	12.70	15.07
Ballia	43.58	47.90	38.80	32.17	4.32	4.03	13.30	15.90
Ghazipur	51.52	58.08	30.52	21.16	4.51	4.82	13.45	15.94
Varanasi	33.26	34.89	25.14	17.40	12.21	16.47	29.39	31.24
Mirzapur	38.86	42.14	40.88	29.60	3.63	7.52	16.63	20.74

TABLE 4.5 (*Contd.*)

1	2	3	4	5	6	7	8	9
Bihar								
Patna	32.61	28.32	34.54	30.73	3.45	4.66	29.40	36.29
Nalanda	40.91	40.18	42.56	41.53	4.03	3.20	12.50	15.09
Nawada	48.57	49.63	38.07	36.00	2.96	2.90	10.40	11.47
Gaya	43.31	43.00	40.06	38.93	3.24	3.25	13.39	14.82
Aurangabad	46.15	46.48	38.52	35.63	3.44	3.87	11.89	14.02
Rohtas	40.94	44.17	39.66	36.76	2.88	3.01	16.52	16.06
Bhojpur	40.17	42.55	41.41	37.53	3.28	3.42	15.14	16.50
Saran	51.69	54.01	33.04	28.03	1.99	3.44	13.28	14.52
Siwan	56.66	62.11	30.67	22.33	1.79	2.36	10.88	13.20
Gopalganj	54.34	63.78	36.89	25.77	1.40	1.85	7.37	8.60
Paschim Champaran	33.58	36.38	55.13	51.36	1.79	1.86	9.50	10.40
Purab Champaran	37.35	42.97	53.53	45.77	1.46	1.87	7.66	9.39
Sitamarhi	37.27	39.59	53.41	48.71	1.73	2.25	7.59	9.45
Muzaffarpur	38.16	40.12	47.33	41.04	1.76	2.50	12.75	16.34
Vaishali	50.47	49.36	36.27	34.09	2.74	3.12	10.52	13.43
Begusarai	34.22	32.98	49.40	47.21	2.64	2.91	13.74	16.90

Samastipur	39.28	40.54	46.78	42.11	2.27	3.38	11.67	13.97
Darbhanga	35.55	37.40	49.58	44.35	2.27	3.24	12.60	15.01
Madhubani	38.74	42.54	51.81	45.92	2.10	3.03	7.35	8.51
Saharsa	42.18	41.05	48.32	48.30	1.29	1.58	8.21	9.07
Purnia	43.59	36.50	43.87	51.30	1.67	1.65	10.87	10.55
Katihar	40.05	34.25	43.14	49.22	2.02	2.37	14.79	14.16
Monghyr	38.35	41.25	43.79	40.00	3.46	3.21	14.40	15.54
Bhagalpur	36.01	38.69	46.38	41.33	3.93	5.32	13.68	14.66
Middle Ganga Plain	44.68	47.21	39.57	34.46	3,12	3.81	12.63	14.69

From : Appendix II of Paper 3 of 1981, Workers and Non-workers, Census of India, Series 1, 1981.

In fact, the Ganga, the Ghaghara, the Rapti, the Gomati, the Burhi Gandak and the Son have all functioned, more or less, in these diverse roles and have acted as centripetal forces for the establishment of settlements and more so for urban centres (Fig. 4.4). The latter have increased in number and magnitude in the last two decades. During the medieval period, many ancient towns (like Vaishali, Mithila, Rajgir, etc.) decayed mostly owing to political vicissitudes, or loss of economic hinges. Sometimes new towns besides the ancient ones, caused the decay of the latter. Sultanpur beside Kushbhawanpur, Mirzapur beside Vindhyachal are cases in point.

A number of towns grew as trade centres or headquarters of dynasties of Delhi Sultanate or other ruling bodies. Some were also developed in memory of royal visits while many grew as citadels of local chieftains or 'Jagirdars'. In the ancient and medieval times and even in early part of the modern period, rivers, mostly unbridged as they were, gave rise to towns on both of their banks because of the break of the overload traffic on either side. During the modern period, however, there has been a change in urbanizing factors, and economic and servicing factors have asserted more. Not only new towns have multiplied but modification and revival of earlier ones have also taken place. Towns have grown up on the cross roads, bridged river sites, railway junctions, and other river towns. The recently increased tempo of economic activity, however, has been generally effective in urban development.

Contact-point locations of diverse physical terrain with different economic attributes and products, like other break-of-bulk points, have also provided significant sites for urban development. Along the northern Tarai and plain zone or the southern Peninsular Upland and Ganga Plain contact zones have developed a series of towns. Chunar, Bhabhua, Sasaram, Dehri, Sheghati, Jamui, Jhajha, Kharagpur and Banka are some of the examples of towns of this type. There are altogether 14 urban agglomerations and Class I cities situated in the Bihar Plains. Seven of these fourteen cities are situated along the Ganga river.

Different aspects of population structures pertaining to these cities, viz., total populations as recorded in 1981 Census,

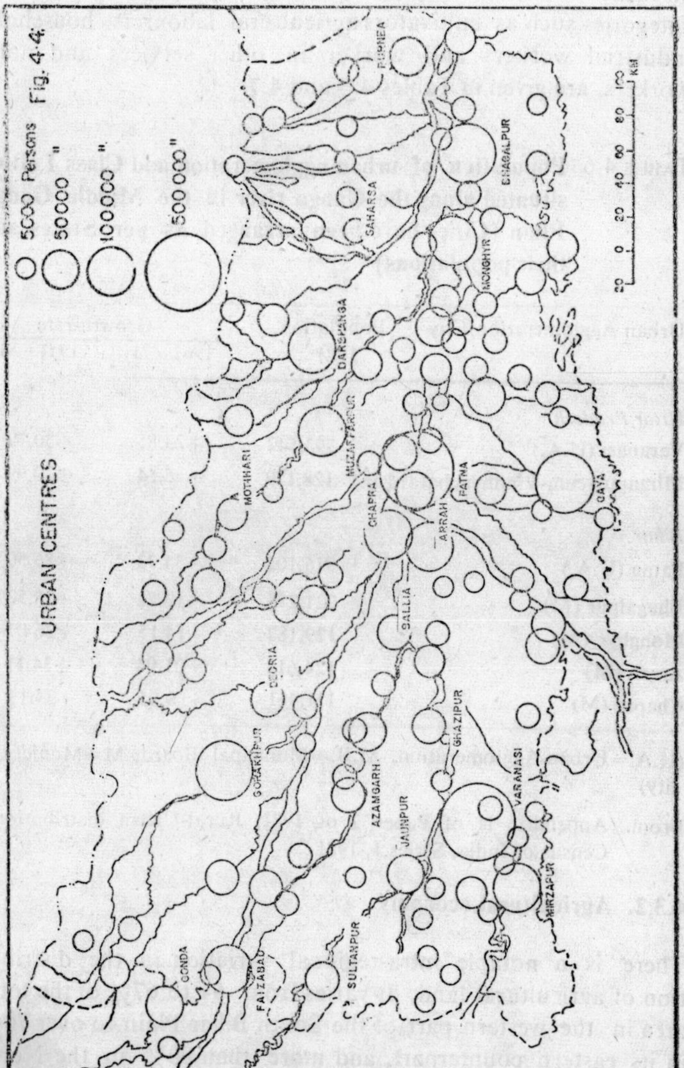
URBAN CENTRES
Fig. 4.4
5000 Persons
50000 "
100000 "
500000 "
KM
PURNEA
SAHARSA
BHAGALPUR
MONGHYR
DARBHANGA
MUZAFFARPUR
MOTIHARI
CHAPRA
PATNA
ARRAH
GAYA
BALLIA
DEORIA
GORAKHPUR
GHAZIPUR
AZAMGARH
VARANASI
JAUNPUR
MIRZAPUR
SULTANPUR
FAIZABAD
GONDA

growth rates for two decennials 1961-71 and 1971-81, total main workers and their percentage to total population and also the distribution of this working population in different categories such as cultivators agricultural labourers, household industrial workers and workers in other services and non-workers, are given in Tables 4.6 and 4.7.

TABLE 4.6 **Population of urban agglomeration and Class I cities situated along the Ganga river in the Middle Ganga Plain (Cities have been arranged as per States and their populations)**

Urban Agglomeration/City	Population, 1981	Growth rate 1961—71	Growth rate 1971—81
Uttar Pradesh			
Varanasi (U.A.)	793,542	+23.85	+30.79
Mirzapur-cum-Vindhyachal (M.B.)	128,179	+ 5.84	+20.99
Bihar			
Patna (U.A.)	916,102	+34.73	+86.50
Bhagalpur (M)	221,276	+19.71	+28.50
Monghyr (M)	129,187	+14.15	+26.07
Arrah (M)	124,614	+21.04	+34.11
Chapra (M)	111,461	+ 9.95	+34.13

(U.A.=Urban Agglomeration, M.B.=Municipal Board, M=Municipality)

From : Appendix II of Paper 2 of 1981, Rural-Urban distribution, Census of India, Series 1, 1981.

4.3.2. Agricultural economy

There is a notable intra-regional variation in the distribution of agricultural land. It varies from 64 to 67% of the total area in the western part of the South Bihar Plain to over 70% in its eastern counterpart, and more than 80% in the North Bihar Plain except in Purnea (77%). In the Eastern U.P. Plain also more than three-fourths of the total land is agricultural in most of the districts except in the 'usar'-infested western part

TABLE 4.7 **Distribution of population by workers and non-workers in the urban ggglomeration and Class I cities situated along the Ganga river in the Middle Ganga Plain (The cities have been arranged as per States and their populations)**

Urban Agglomeration/City	Main workers						
	Total workers	Workers as per cent to total population	Total culti- vators	Agricultural labourers	Household industry	Other workers	Non-workers
Uttar Pradesh							
Varanasi (U.A.)	214,627	27.05	4289	3136	62,122	1,47,045	576,950
Mirzapur-cum-Vindhyachal (M.B.)	36,542	28.51	842	1206	6,687	27,807	91,637
Bihar							
Patna (U.A.)	237,162	25.89	11086	17865	16,513	1,97,716	672,922
Bhagalpur (M)	52,268	23.62	2178	3521	9,800	37,274	168,503
Monghyr (M)	26,761	20.71	1229	2141	968	22,822	102,027
Arrah (M)	28,402	22.79	2102	3008	1,483	22,070	95,951
Chapra (M)	25,648	23.01	2410	4248	2,264	17,832	84,707

(U.A.=Urban Agglomeration, M.B.=Municipal Board, M=Municipality

From : Appendix IV of Paper 3 of 1981—Workers and Non-workers, Census of India, Series 1, 1981.

of the Doab (below 70%) and forested areas of Gonda and Varanasi districts. The percentage increases from south to north and west to east.

In the Western South Bihar Plain the net sown area ranges from 87% to 92% while it ranges between 69% (Purnea) and 88% (Muzaffarpur) in North Bihar Plain and 77% to 81% in Eastern South Bihar Plain. Fallow land percentages are the highest in North Bihar Plain and South Bihar Plain, the highest being in Purnea (over 29%) and below 7% in Western South Bihar Plain. The same trend is found in the distribution of cultivable land which is lowest in Western South Bihar Plain but goes as high as 8.25% (Monghyr) in East South Bihar Plain and over 11% (Saharsa) in parts of North Bihar. It is low in Purnea (2.28%) and Muzaffarpur (2.92%).

Of the total agricultural land, the net sown area percentage is generally higher in East U.P. Plain than in Bihar Plain, the range being between 84 and 94%. The net sown area percentage increases from west to east, coinciding with the incidence of rainfall, and in consonance with the same there is a north to south decreasing trend. Culturable waste percentages are correspondingly low, the overall range being 2.1 (Deoria) to 7.3 (Sultanpur). The current fallow also has a similar trend.

4.3.2.1. Irrigation

Irrigation plays a significant role in the farming operations and agricultural land use of the region. Seasonal concentration and ill-distribution of rainfall, both in space and time, and other monsoonal vagaries necessitate supplemental irrigation for not only annual crops like sugarcane or zaid crops which are grown in the precarious summer, but also for higher yields of the 'kharif' and 'rabi' crops. Now that the farmers are picking up modern farming techniques, need of the availability of adequate amount of water to the fields at opportune time cannot be over-emphasized.

Of the net sown area, about 33% in the Bihar Plains and over 37% in East U.P. Plain are irrigated. In general, the areas of relatively higher rainfall and lower variability have lower percentage of irrigated area. In the more rainy N. Bihar Plain less than 7% of the net sown area is irrigated while the less rainy S. Bihar Plain has as much as 60% of area under irrigation While the Kosi Plain, where rainfall is the highest and variability the least, only 4.5% of the net sown area is irrigated. In the eastern part of the S. Bihar Plain, where rainfall ranges from 100 to 120 cm and where variability index is 20, about 75% of the net sown area is under irrigation. Also, over 1/4th of the sown area of the Anga Plain is irrigated.

In the East U.P. Plain, unlike most of the Bihar Plain, wells and tubewells are the most significant sources of irrigation, and nowhere except in Basti and Gorakhpur in the Saryupar, do they irrigate less than 65% of the total irrigated area.

The power-driven tubewells bid good promise on the region as they can be sited in compact, well-drained blocks of 200-300 hectares or so. They can serve the area most without interfering with irrigation from tanks, 'jhils' and surface wells in the low-lying areas. However, so far, there has been no detailed scientific ground water survey, and there is a feverish activity in tapping ground water through diesel or electric pumps from different levels, often causing drying-up of the country wells leading sometimes to scarcity of even drinking water in the rural areas. In this region, impounding of rainwater in surface reservoirs or 'Arhars' by throwing embankments across the small rivulets will go a long way in assuring irrigation by ponds and wells, at the same time maintaining the water table of the village wells.

4.3.2.2. *Cropping pattern*

The cropping pattern is typical of an underdeveloped agricultural economy with most of the cultivated area (about 90%) devoted to subsistence food crops, while the cash crops or industrial raw materials like sugarcane (4.8%) or oilseeds (1.4%) have negligible coverage. With an increasing use of the HYV seeds and modern techniques of production and growing facilities for irrigation, storage, marketing and transportation, some traditional subsistence crops such as rice, wheat or maize have become 'money' crops at least for the well-to-do farmers, and are competing with sugarcane, jute, etc. Jute has been recently introduced in parts of the humid zone, particularly in the Tarai and Purnea districts.

Rice is the dominant crop in the East U.P. Plain claiming over 32% of the cultivated area and is followed by gram (about 16%), wheat (11.5%), barley (8.0%) and maize (5.8%). About 1/6th of the cropped area is given to various pulses other than gram, like Arhar, Mung, Masur, etc., and over 4% to vegetables, zaid crop, etc.

Wheat, the most important rabi crop, is concentrated in the rich, moisture-retentive loam silts of N. Bihar Plain west of the Baghmati in S. Darbhanga and contiguous N. Monghyr and the irrigated Gaya and Shahabad districts in the Western S. Bihar Plain.

Maize is the third major crop concentrated in the well-drained, fertile sandy loam tracts, south and west of the Burhi Gandak and parts of Saharsa, Purnea and Monghyr. Barley, next to wheat among 'rabi' crops, covers NW Saran, Champaran and Muzaffarpur in N. Bihar Plain. It is invariably mixed with gram. Pulses are important both as main and cash crops, and together stand second to rice. Gram alone covers 1/5th of all pulses or about 5% of the total cropped area of Bihar Plain.

Oilseeds are widely grown as cash crops, and Purnea, Shahabad, Gaya and Champaran are well-known for their cultivation. Sugarcane thrives in the 'bhat' soils of the south-western and western parts of N. Bihar Plain, contiguous to the East U.P. sugarcane belt and feeds the local sugar factories. It replaced indigo cultivation in the thirties.

Jute, although a recent entrant in the crop economy stands next to rice in Purnea and Saharsa which together produce 80% of jute in Bihar.

4.3.2.3 Crop combinations

There are two important seasonal crop-systems—'kharif' and 'rabi', not so well marked as in the Upper Ganga Plain. Although 'bhadai' and 'zaid' (summer) seasons do operate here, they are not so strongly distinguishable as in the perhumid Bengal. The Bhadai crops (early rice, maize, inferior millets like 'sawan', 'khodo', 'marua', etc.) and 'Agahani' crops such as millets (jowar, bajra), arhar and mung are grown with partial inter-culture. Such inter-culture crops not only thrive by drawing different properties to different extent in the fields because of their differential agronomical needs but they are also a check to the monsoonal vagaries, providing some security against the mischances of the season. The various combinations of crops practiced in the region are given in Fig. 4.5.

4.3.3. Fisheries

As mentioned in the previous chapter, a detailed investigation on the fisheries of the Ganga river system have been conducted during the years 1958-66 by the Allahabad sub-station of the CIFRI. The investigations comprised total enumeration of daily arrivals of fish at Varanasi, Buxar, Ballia, Patna and Bhagalpur. The annual market arrivals at the above stated centres are shown in Table 4.8.

4.3.4. Industrial economy

On the industrial map of India, sugar excepted, the Middle Ganga Plain appears to be a void with its few individual prized stars—Sahupuri (Chemicals), Maruadih (diesel locomotive), Gorakhpur (fertilizer) in East U.P. Plain and Dalmianagar (a complex of cement, chemicals, sugar and vegetable oil) in S. Bihar and Barauni (Petro-chemicals, fertilizer, Bata shoe factory and McDowell Distillery at Mokamah) in N. Bihar Plain. The rather larger railway workshops of Gorakhpur and Jamalpur also are important (Fig. 4.6).

Among the household industries, the textiles and other agro-based industries together account for more than 50% of total employment.

Among the factory industries also agro-group and textiles predominate. Agro-group takes the lead in 14 districts employing over 40% of the total factory workers.

The sugar factories numbering 55 (1964) employ 38,000 persons. A large part of the cane grown is converted into 'Gur' and 'Khandsari' more particularly in U.P. The maximum concentration of the industry is in the 'Bhat' soil zone. Outside the main belt, there are only six sugar mills, located south of the Ganga in Gaya (3) and Shahabad and Patna (each with one) and only one at Shahganj in the Ganga-Ghaghara Doab. Some cooperative sugar factories have also been established as

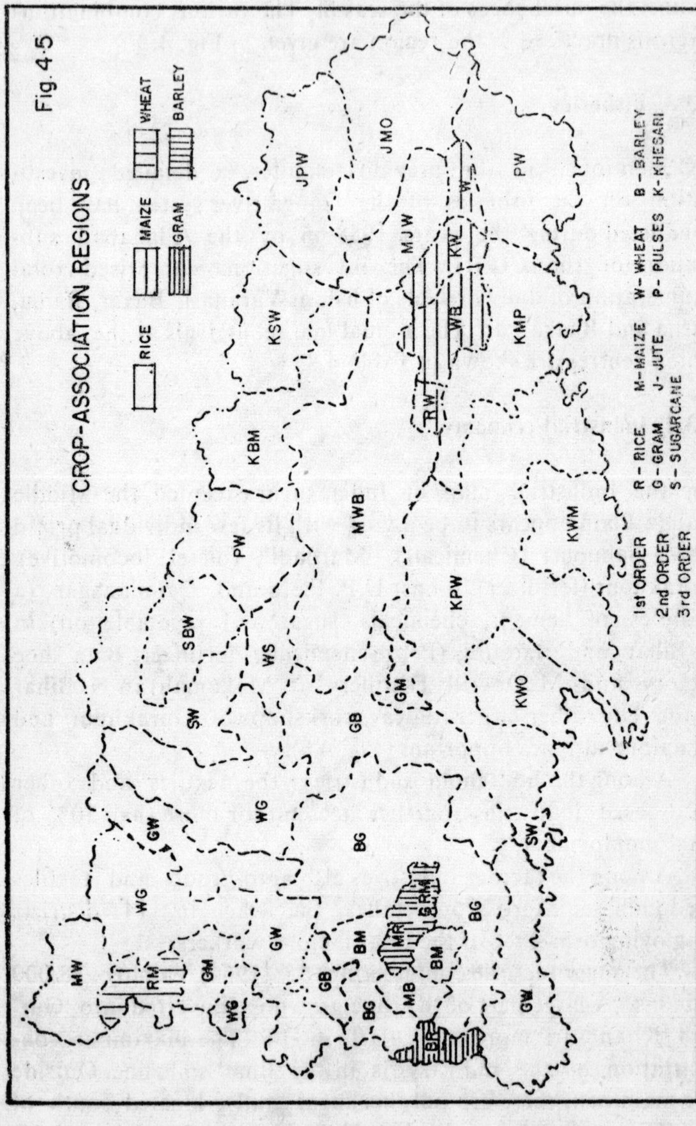

CROP-ASSOCIATION REGIONS

Fig. 4·5

TABLE 4.8 **Fish landings (t) at different centres situated along the Ganga river system in the Middle Ganga Plain during the period 1955-56 to 1965-66**

Year	Name of the centre				
	Varanasi	Buxar	Ballia	Patna	Bhagalpur
1958-59	46.71	108.62	45.38	54.60	41.51
1959-60	47.16	71.77	50.24	40.11	59.88
1960-61	69.38	54.27	84.40	96.39	85.98
1961-62	71.14	—	51.27	106.31	120.55
1962-63	105.36	—	72.37	129.28	98.36
1963-64	75.18	32.58	114.23	115.38	77.65
1964-65	71.02	31.63	—	76.96	76.35
1965-66	109.92	56.12	—	114.95	94.92
Average	74.48	59.17	69.64	91.85	81.93

From : Jhingran *et al.*, 1970.

at Dobhi (Jaubour), Aurai (Varanasi), Ghosi (Azamgarh), etc. In the last two decades, the number of sugar factories may have increased but exact figures of the increase are not available.

Other agro-group industries comprise rice, flour, dal, oil-crushing, fruit and fish-canning and daily products, etc. They are widespread being located in small towns and bazars in general, with significant clustering in larger centres.

Cotton textile units including power-borne centres are located in Phulwari (near Patna), Buxar, Gaya, Varanasi, Mirzapur, Mau, Akbarpur, Tanda, Khalidabad, Madhubani, Bihar Sharif, Mubarakpur, etc., most of them being traditionally important handloom centres. Mirzapur, Varanasi and Bhadohi are important woollen carpet centres. Jute industry, a new entrant, is located at Katihar, Samastipur and Sanjanwa (Gorakhpur). Varanasi is famous for 'Banarasi Silk' with more than a dozen large units, while Bhagalpur, where one more unit has been recently added with Japanese collaboration, is important for 'Tassar' silk.

Dalmianagar on the Son at the tip of the Rohtas Plateau in Shahabad district is a heterogeneous industrial complex with cement, asbestos, chemicals, paper and pulp, card-board and

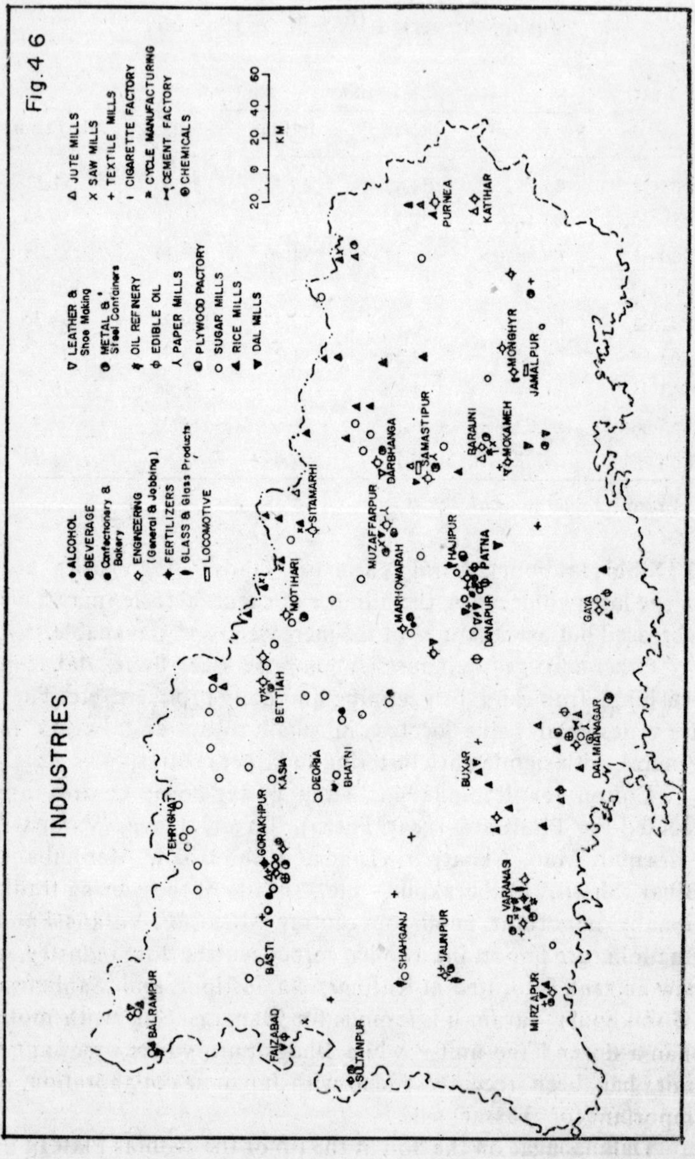
INDUSTRIES
Fig. 4.6
ALCOHOL
BEVERAGE
Confectionary & Bakery
ENGINEERING (General & Jobbing)
FERTILIZERS
GLASS & Glass Products
LOCOMOTIVE
LEATHER & Shoe Making
METAL & Steel Containers
OIL REFINERY
EDIBLE OIL
PAPER MILLS
PLYWOOD FACTORY
SUGAR MILLS
RICE MILLS
DAL MILLS
JUTE MILLS
SAW MILLS
TEXTILE MILLS
CIGARETTE FACTORY
CYCLE MANUFACTURING
CEMENT FACTORY
CHEMICALS
20 0 20 40 60
KM
BALRAMPUR
TEHRIGHAT
GORAKHPUR
BASTI
FAIZABAD
SULTANPUR
SHAHGANJ
JAUNPUR
MIRZAPUR
VARANASI
BUXAR
KASIA
DEORIA
BHATNI
BETTIAH
MOTIHARI
MARHOWARAH
MUZAFFARPUR
SITAMARHI
DARBHANGA
SAMASTIPUR
HAJIPUR
PATNA
DANAPUR
GAYA
DALMIANAGAR
BARAUNI
MOKAMEH
MONGHYR
JAMALPUR
PURNEA
KATIHAR

plywood, sugar, vegetable ghee factories developed since 1933. Barauni has a large oil-refinery and associated petro-chemical complex based on Assam crude piped to this place. It also has a fertilizer complex. Bata shoe factory and McDowell Distillery at Mokamah (Barauni) are some of the industries worth mention in the region.

The district headquarters and large cities like Varanasi, Patna, Gorakhpur, Bhagalpur, Gaya, Muzaffarpur, Darbhanga and Mirzapur, some with industrial estates, have an industrial status in their own right, having a host of textile, food processing, light engineering and foundry units, chemical, electrical and various other consumer goods units, as well as printing and publishing units. Varanasi is developing apart from its traditional lines, due to some important industrial units such as the large Diesel Locomotive (Maruadih), chemicals (Sahupuri across the Ganga), glass (Ramnagar), Asia Cycle Works, and dozens of various industrial units between Varanasi and Moghalsarai along the Grand Trunk Road.

4.3.5. Transport and communication

The basic structure of the transportation system of the region consists of a network of roads of diverse nature and capacity, railways operated mostly by the Northern and Eastern (both Broad gauge) and North-Eastern (Meter gauge) Railways, and a rather incomplete net of airways. Rivers are navigated by country boats of various sizes and ferry steamer services to Saigon also passes through the region with important link with Kathmandu. This highway has provided a great fillip and a spurt for road and bridge construction in the region. Being itself a densely peopled region with heavy demands for communications and also located midway in the different regions of the Great Plains, east-west and Peninsular parts and Nepal, the regional transport arteries always stand on the nerve for movement of enormous quantities of food and industrial raw materials and goods as well as heavy passenger traffic. Intra-regional transport always suffers from the enormity of the bridge traffic emphasizing a need for a huge expansion of the facility.

Three main railways run through the region—two broad

gauges, one running south of the Ganga via Mirzapur and the other north of it up to Varanasi. Further, both run towards Calcutta along different routes from Mogalsarai, the biggest railway yard of India. The third railway is a metre gauge but large net serving the Doab, the Saryupar and N. Bihar, linking all these areas to Assam. The region has direct linkage to Delhi, Calcutta and Bombay, and via Chunar-Chopan-Garhwa railway to the vast interior of the immediate Peninsular lands. From the map (Fig. 4.7) it is obvious that the general outline of the railway net is almost complete except a few gaps, particularly in the Kosi tract of N. Bihar and the Saryupar Plain. These areas are more riverine, have medium to high drainage texture, and suffer from severe floods and shiftings of the river channels which impede the development of transport arteries, particularly the railways. Because of the shift of the Kosi, the railway lines from Forbesganj to Anchra Ghat, Anchra Ghat to Nirmali, and Nirmali to Supaul were dismantled and could not be reconstructed. The Bagha-Chhitauni link joining the Saryupar and the northern part of the N. Bihar Plain could not be re-established since 1924 when the Bagha bridge collapsed during the Gandak floods.

The different rail-gauges and single track system on many of them put extra burden on the existing arteries and the break-off bulk points created as such hold the traffic to the point of inefficiency. In North Bihar and also in the Saryupar, railway and roads run parallel to one another as well as to the drainage lines because of bridging difficulty and costs, and also interference with drainage, all resulting in long circuitous and time-consuming routes from one town to another.

With the advent of railways and roads, the importance of the rivers for navigation fell. Since the closure of Calcutta-Patna steamer service on the Ganga, the water transport is limited to ferry-steamer service, ferry-boats and countryboat services.

The vast tract of the eastern Doab has high density of surfaced roads with over 1 km of road per 10 km^2. The tract is formed of two blocks—(i) districts of Azamgarh, Jaunpur, Ballia and parts of Allahabad, Pratapgarh, Gorakhpur, Deoria, Ghazipur and Varanasi in the East U.P. Plain, and (ii) the compact zone of western portion of Patna, northern Gaya and

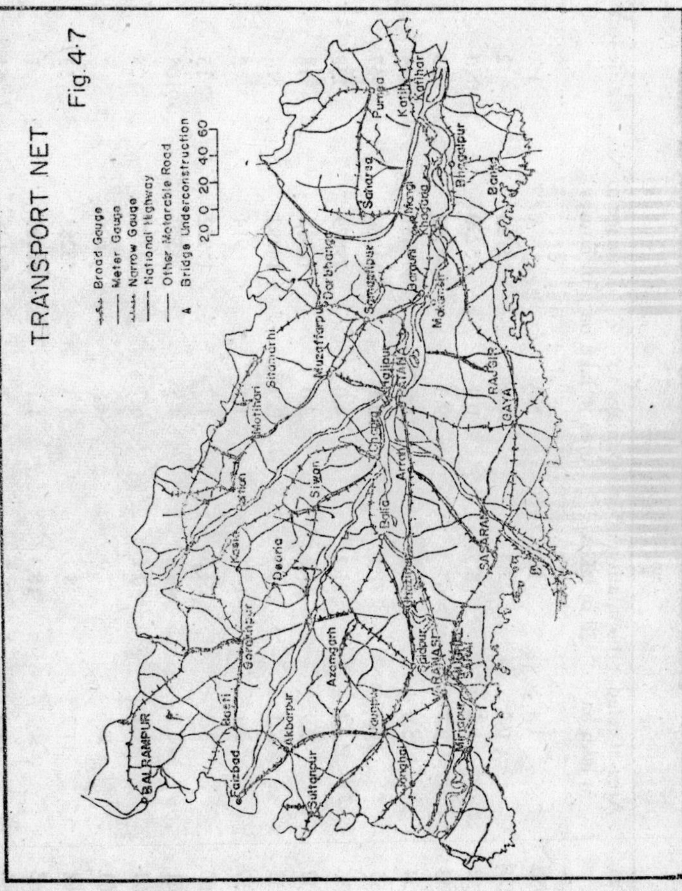
TRANSPORT NET
Fig. 4·7
Broad Gauge
Meter Gauge
Narrow Gauge
National Highway
Other Motorable Road
Bridge Underconstruction
20 0 20 40 60

TABLE 4.9 **Flood damage during 1953 to 1981 in the State of Bihar**

Sl. No.	Year	Area affected in lakh ha	Population affected in lakhs	Damage to area in lakh ha	Damage to No. of houses	Cattle lost	Human lives lost in Nos.
1.	1950	1.4	NA	NA	NA	150	NA
2.	1951	0.4	2.4	Neg.	5120	2	Nil
3.	1952	1.3	8.5	0.4	5230	Nil	Nil
4.	1953	9.7	67.7	9.3	184300	33	5
5.	1954	25.0	76.7	15.9	179451	1944	63
6.	1955	17.7	71.1	10.1	11356	57	44
7.	1956	13.2	31.2	4.0	79879	138	32
8.	1957	7.9	17.8	3.0	18812	18	7
9.	1958	7.1	23.7	2.9	14393	Nil	6
10.	1959	2.1	10.1	1.5	914	2	Nil
11.	1960	13.2	24.5	1.7	15952	15	7
12.	1961	12.7	46.8	4.5	31570	10105	551
13.	1962	11.1	40.1	6.7	64397	25	20
14.	1963	2.7	7.0	Neg.	606	2	Nil
15.	1964	11.2	38.1	5.3	53285	52	3
16.	1965	4.3	18.3	1.1	86185	1	2
17.	1966	15.3	58.7	11.2	56125	64	3038

18.	1967	12.5	64.3	7.3	77699	68	23
19.	1968	7.3	35.6	1.9	76503	7821	13
20.	1969	9.7	41.7	5.1	69247	18	31
21.	1970	9.3	38.9	3.8	16989	1855	2
22.	1971	42.6	217.1	15.5	786000	426	23
23.	1972	2.2	9.0	0.2	2225	Nil	Nil
24.	1973	7.3	31.4	3.5	8819	6	8
25.	1974	31.4	164.3	17.5	432000	321	80
26.	1975	23.1	132.4	11.7	320311	1323	148
27.	1976	29.9	136.4	8.5	494060	4518	186
28.	1977	11.5	30.8	2.3	19741	66	16
29.	1978	23.7	120.6	9.7	310003	466	177
30.	1979	8.1	37.4	2.7	27816	4	14
31.	1980	19.2	74.5	10.0	81015	42	57
32.	1981	12.6	69.5	7.7	75776	11	18
	Total	406.7	17461	185.0	3705479	29553	1603
	Maximum	42.6 (1971)	217.1 (1971)	17.5 (1974)	786000 (1971)	10105 (1961)	551 (1961)
	Average	12.7	56.3	6.0	119532	924	52

NA=Not Available, NR=Not Reported, Neg=Negligible

From : Central Water Commission, New Delhi.

north-eastern of Saran and Muzaffarpur districts. The marshy and more riverine tract covering the area north of Basti-Gorakhpur-Hata line running through northern Saran, S. Champaran and northern one-third of Darbhanga and the entire Saharsa-Purnea fall below 0.8 km per 10 km^2; in fact, Purnea and Champaran have mostly below 0.6 km roads per 10 km^2.

Varanasi and Patna are on the air map of India with flights to Kathmandu (Nepal) also. However, there is no direct air connection to the south.

4.4. PROBLEM IN THE REGION

The above discussion sets forth many problems of the Middle Ganga Plain and its people.

The problems are of two kinds—(i) Those created by natural causes, and (ii) those created by man.

4.4.1. Due to natural causes

Though these problems are created by the fury of nature, the ill-effects are often accelerated or aggravated by human agency. As such, the two types of problems are hardly different.

The region is richly fed by rivers. As a consequence, it has to face the ugly side of the riverine system, i.e., floods and erosion. The land in the region has borne a great burden of these. There are almost no forests, or grasslands or pastures, no grazing lands, and methodical soil management is missing. As a result, recurrent floods is a permanent feature in the region. Since all the river s ystems in Bihar are concentrated in this region, the statement of flood damages of Bihar for the years 1953-81 are given here as a representative of the region (Tables 4.9 and 4.10). The average total damage to crops, houses and public utilities due to floods tunes to Rs. 5,649.3 lakhs with the maximum damage amassing to Rs. 38,183.6 lakhs reported for the year 1974. The maximum damage reported in terms of area, population, crops, houses, public utilities, etc., was in the year 1971 and again in 1974. Highest

TABLE 4.10 **Flood damage in lakhs of rupees during 1953 to 1981 in the State of Bihar**

Sl. No.	Year	Crops value	Houses value	Public utilities	Total damage
1	2	3	4	5	6
1.	1950	100.00	150.0	7.0	257.0
2.	1951	2.0	4.0	NA	6.0
3.	1952	30.0	50.0	2.0	82.0
4.	1953	2900.0	600.0	150.0	3650.0
5.	1954	1500.0	500.0	150.0	2150.0
6.	1955	2030.0	112.0	20.5	2162.5
7.	1956	522.0	51.0	NA	573.0
8.	1957	NA	NA	NA	NA
9.	1958	319.8	NA	1.0	320.8
10.	1959	NA	NA	NA	NA
11.	1960	NA	NA	0.6	0.6
12.	1961	1325.0	NA	NA	1325.0
13.	1962	957.0	51.6	4.5	1013.1
14.	1963	75.8	0.6	NA	76.4
15.	1964	1108.8	47.5	171.0	1327.3
16.	1965	180.4	NA	NA	180.4
17.	1966	3900.0	NA	NA	3900.0
18.	1967	1934.0	27.0	6.0	1967.0
19.	1968	1014.0	1.0	1.0	1016.0
20.	1969	2880.0	NA	720.0	3600.0
21.	1970	1474.0	32.0	1.7	1507.7
22.	1971	14391.2	2930.5	4422.1	21743.8
23.	1972	162.0	7.2	1.0	170.2
24.	1973	1970.6	34.6	0.9	2006.1
25.	1974	29495.0	5287.6	3401.0	38183.6
26.	1975	13655.3	969.2	11951.8	26576.3
27.	1976	9463.9	3003.5	8132.0	20599.4
28.	1977	1178.8	45.4	4.4	1228.6
29.	1978	9460.3	912.2	7513.0	17885.5
30.	1979	1901.5	103.4	14.0	2018.0

TABLE 4.10 (*Contd.*)

1	2	3	4	5	6
31.	1980	NR	372.3	5401.0	5773.2
32.	1981	7213.2	397.7	298.6	7907.5
	Total	111144.6	15688.3	42375.1	169208.00
	Maximum	29495.0 (1974)	5287.6 (1974)	11951.8 (1975)	38183.6 (1974)
	Average	3969.5	653.7	1765.6	5640.3

NA—Not Available, NR—Not Reported, Neg.—Negligible.
From : Central Water Commission, New Delhi.

number of cattle head and human lives lost were reported in the year 1961 with the respective numbers being 10,105 and 551 (Table 4.10).

4.4.2. Those created by man

The densely populated region faces the enormous problem, of population explosion, with all its fallout-shortage of food, malnutrition, disease, health, education, employment, etc. There is also the land problem; diminishing land-holding or per capita land, that is getting more acute everyday due to the lack of alternative channels of livelihood.

The region, though mainly rural, has shown slow, but definite trend towards urbanization. There is an increase of about 2.57% in the urban population and 2.58% in the workers of the household industry and services. Industrialization of the region is on an increase, and hence the offshoot of industrialization, i.e., pollution, has entered the region. The problem of pollution is mostly confined to the urban agglomeration and Class I cities which are more industrialized than the cities belonging to other classes. Hence, the problem becomes more localized but its impact is overspread to a larger area, beyond its occurrence. However, the presence of the problem in smaller cities cannot be ruled out.

Since the region is naturally endowed with one of the best developed river systems in the world, the industries as well as the domestic bodies, more often than not, dump their wastes many a times untreated, in these rivers affecting the quality of the rivers. Since Ganga is the main stream of the region, it is the centrally focussed river in this respect. There are few reports about the human settlements and its impact on the Ganga river. Those available are given below.

As stated earlier, the region abounds one of the oldest living city of the country and perhaps of the world, i.e., Varanasi. Size of this city has not however, remained the same as in the bygone days, as a result of which there is great pressure on the river Ganga in the form of increased domestic and industrial wastes that are dumped into the river everyday. Consequently, quality of river water gets progressively deteriorated. In his studies of algal communities of the Ganga river at Varanasi, Rai (1978) reported the presence of *Stigeoclonium* and *Schizomeris* in some areas of the river (Rajghat). These are indicators of organic pollution of water. He also revealed that the species like *Cylindrotheca*, *Gyrosigma* reported to be the inhabitants of less polluted waters, were not present in that region. In another study, Bhargava (1981a) has prepared some water quality indices (WQI) for the Ganga river at Varanasi, on the basis of his observations about the physico-chemical characteristics of the river (Table 4.11). These indicate that the quality of water at the upstream of Varanasi is good but that at downstream is poor to satisfactory for various

TABLE 4.11 **Observations along Ganga in the Middle Ganga Plain, during summer period**

	Varanasi	
	Upstream	Downstream
Temperature °C	27	31
DLP cm	73	78
Do mg/l	6.4	5.8
BOD mg/l	4.0	10.0
Conductivity m mhos/cm	494	515
Chlorides mg/l	34	34
Hardness as $CaCO_3$ mg/l	154	158
Coliform MPN no/100 ml	49	2400

From : Bhargava, 1981a.

human uses (Table 4.12). He further points out the fact that the quality of river can be improved upon, only by treatment of wastes (domestic/industrial) before being released into the river.

TABLE 4.12 **Water Quality Indices (WQI) and classification of Ganga for different uses in the Middle Ganga Plain during summer period**

Beneficial uses	Varanasi			
	Upstream		Downstream	
Without treatment drinking, bathing	75	II	33	IV
Public water supplies	83	II	47	III
Agriculture	71	II	70	II
Industrial	71	II	72	II
Fish culture	74	II	53	III
Overall WQI	75	II	55	III

I=≮90—Excellent, II=65—89—Good, III=35—64—Satisfactory, IV=11—34—Poor, V=<10—unacceptable.

From : Bhargava, 1981a.

Besides discharging of wastes in the river, the Hindu belief of getting 'Moksha' by dumping dead bodies into the 'Holy Ganga' also poses great danger to the 'Holiness' of the river. It is estimated that around 30,000 bodies are burnt along two 'ghats' at Varanasi every year. Between 1/2 and 2 tons of wood is needed to burn each dead body. All the resultant ash goes into the river. Besides, innumerable unburnt bodies are also let off into the river 'to attain Moksha', adding to the organic detritus of the river, causing severe organic pollution in the form of increased coliform bacteria in the river (Table 4.11). The number of both the types of disposal may be on an increase with improved transportation facilities. However, an intensive study is required to assess the damage caused by these activities, to arrive at reliable conclusions.

In the lower stretches of the Ganga in the region, a detailed limnological survey and assessment of impact of human activities on the river, was carried out at the departments of Botany and Zoology of the Bhagalpur University, Bhagalpur. Some of the important findings of the study are quoted below (Bilgrami & Datta Munshi, 1979).

4.4.2.1. Area of study and sources of pollution in the river Ganga from Barauni to Farakka

The range under investigation was about 256 km from Barauni to Farakka. Seven centres were selected for regular survey, viz., Barauni, Monghyr, Sultanganj, Bhagalpur, Sahebganj, Rajmahal and Farakka.

Important sources of pollution in the stretch of the river under study include discharge of effluents from industrial complexes, sewage, and waste materials of town. Human activities like washing, bathing, dumping of dead bodies and remains from cremation grounds, excessive use of insecticides, pesticides and fertilizers, excessive navigation and transportation and brick-making occur on river banks. All these activities exert heavy strain on the river. Maximum industrial pollution and contamination of the river takes place near Mokamah bridge where effluents from a number of factories are discharged. At this site, Bata Shoe Factory and McDowell Distillery are situated on the southern bank (Mokamah) of the river. An Oil Refinery, Thermal Power Station and a Fertilizer Factory are located on the northern side of the river (Barauni).

Physico-chemical and biological characters of the river water were studied for a period of 3 years from 1976-79.

4.4.2.2. Physico-chemical characters

Bicarbonate alkalinity, dissolved oxygen, and temperature did not exhibit any notable fluctuations. Degree of turbidity, colour of water and amount of suspended materials however, had a definite relation with each other. Higher turbidity recorded at certain stations (Near Rajendra Bridge and Mixing zone of Barauni Oil Refinery effluent) was obviously due to addition of coloured wastes and suspended particles from the industries.

Except at the above-mentioned sites, turbidity was less during summers and winters, practically at all the stations. The range of fluctuation in pH (7.2-8.2) of water was comparatively narrower at different sites except near the mixing zone of industrial effluents. This value was slightly higher at Barauni in the mixing zone of Barauni Oil Refinery effluent (8.4-8.8) due to excess of alkaline minerals alongwith petroleum wastes of Oil Refinery. At station Rajendra Bridge, the acidity of distillery westes lowered the pH (6.0 6.6) range of river water. This reduction in pH was accompanied by significant rise in free CO_2 (24 ppm to 45 ppm). The maximum value of dissolved O_2 was recorded during winter (7.0-14.5 ppm), which was possibly due to low water temperature during winters which enhanced the oxygen dissolving capacity of water.

Maximum concentration of silicate was estimated during the monsoon (20 25 ppm). This may be attributed to run-off of silicate-containing material from external sources. Depletion in the concentration in summer (9.0-16 ppm) was associated with higher population of phytoplankton (Table 4.13). This indicated that major fraction of silica was used in the formation of wall of diatoms during the summers.

4.4.2.3. Biological characters

The total density of phytoplankton varied with seasons. Its concentration was very high in winters and summer while it was low during the rains. Amongst phytoplankton, Chlorophytes dominated over the Bacillariophytes and Cyanophytes (Table 4.13).

The quantitative and qualitative pattern of Zooplanktonic forms at different sites varied with seasonal fluctuations. It had two peaks of production—in summer (14-34/litre) and in winter (14-26/litre). The population became scarce due to high current and heavy load of soil particles in water during rains. Their number varied from 5-9/litre. Such pattern of variation showed a close relation to rainfall, wind circulation and also the load of pollutants (Table 4.13).

Altogether, fifty macrophytes were collected during the tenure of the survey. It was observed that the number of marshy inhabitants was greater than that of the aquatics.

TABLE 4.13 **Seasonal fluctuations (number/litre) of phytoplankton (P) and zooplankton (Z) at different sites of the river Ganges (1978-79)**

	Sampling stations	Summer		Monsoon		Winter	
		P	Z	P	Z	P	Z
1.	Mokamah	—	34	—	7	—	23
2.	Barauni	4137	17	530	7	1398	20
3.	Monghyr	3513	20	405	8	1326	26
4.	Sultanganj	1484	14	337	8	1107	21
5.	Bhagalpur	2614	27	377	6	1041	19
6.	Sahebganj	1605	25	318	5	870	21
7.	Rajmahal	4720	27	343	8	3091	18
8.	Farakka	2404	31	298	9	2867	14

From : Bilgrami & Datta Munshi, 1979.

Among the submerged forms *Potamogeton indicus* was confined to Farakka while *Vallisneria spiralis* was recorded at Bhagalpur, Monghyr and Kahalgaon. Primary production of free floating and submerged species differed considerably. Biomass and productivity of free floating and submerged species showed an inverse correlation.

Bacteria and Fungi constituted major group of decomposers. Most of the bacteria were of aerogenous, streptococcal and gram negative type. The most probable number (MPN) of coliform bacteria per hundred ml of sample water varied greatly with research sites. Seasonal fluctuations also affected the population of coliform bacteria. Its maximum concentration was in water samples collected from Sultanganj, Bhagalpur and Monghyr sites. Bacterial population is also found to be influenced by season and temperature. Low temperature during winter decreased the concentration, whereas comparatively high temperature was more suitable for their growth (Table 4.14). A pH range (7.5.8.5) and phosphorus range (0.14-0.82 ppm) was found to be suitable for bacterial growth at Monghyr, Sultanganj and Bhagalpur.

In addition to purely aquatic fungi, some semiaquatic fungi and spores of terrestrial forms were isolated in different seasons.

TABLE 4.14 **Variations in the number of coliform bacteria in the river Ganges during 1978-79**

Year 1978/1979	Maximum probable number of coliform bacteria (MPN) per 100 ml water at different sites							
	Barauni	Mokamah	Monghyr	Sultanganj	Bhagalpur	Sahebganj	Raj Mahal	Farakka
May 78	161	28	24	161	162	35	92	40
June 78	173	84	345	345	94	345	141	49
July 78	141	32	212	278	253	48	41	120
Aug 78	212	177	221	542	278	120	84	148
Sept 78	221	40	109	542	109	84	39	39
Oct 78]	345	172	348	426	278	109	79	94
Nov 78	253	120	212	345	177	148	94	48
Dec 78	177	84	148	221	148	120	84	76
Jan 79	141	76	120	175	109	94	76	53
Feb 79	120	79	94	141	94	79	53	49
March 79	94	40	109	221	109	70	40	53
Apr 79]	148	63	177	253	172	84	76	33

From : Bilgrami & Datta Munshi, 1979.

Basidiomycetes and *Mycelia sterila* were poorly represented. Representation of fungi imperfecti was maximum in this range. These were represented by 63 species. Their number declined during the floods which may be attributed to heavy water current, low oxygen contents and high turbidity. Industrial effluents, i.e., wastes from McDowell Distillery, Bata Shoe Factory (Mokamah) and Oil Refinery (Barauni) inhibited the growth of fungi but their concentration was usually high at the spots where sewage from the towns joined the river.

During rainy season, the decomposition rate was generally higher. This was possibly on account of higher bacterial concentration at that time.

During the course of extensive survey of fauna, altogether seventy-two Teleostean species of different taxonomic groups were recorded.

One species of *Rana*, 10 species of tortoise, one species of natrix, 13 species of birds, one mammalian species, i.e. *Platinista gangetica* and seven species of aquatic insects were collected, identified and classified.

The stretch of river from Barauni to Farakka has diverse beds of fishing. The pH (6.4 to 7.4), dissolved O_2 (6.0 to 9.5 ppm), water temperature (29 to 31°C) and overcast sky with gentle breeze are most suitable for spawning. The polluted region of the river near Mokamah-Barauni industrial complex serves as a barrier for upstream movement of fishes and possibly this is one of the major reasons for lack of spawning centre in this area (Table 4.15). The polluted zone also prevents the migration of fish from north to south and vice versa.

4.4.2.4. Impact of human activities on the Ganga from Barauni to Farakka

4.4.2.4.1 Impact of Industries : As mentioned earlier, most of the pollution of the river takes place due to the industrial complex at Mokamah-Barauni. Extent of pollution and its impact caused by the respective industries is given below.

Indian Oil Refinery is a complex of units processing petroleum products. The effluents are discharged in the Ganga

TABLE 4.15 **Spawn collection grounds with some physico chemical factors and average catches in the river Ganges (July-August, 1976)**

Spawn collection grounds	Current speed km/hr	Secchi disc transparency (cm)	Temperature (°C)	pH	Dissolved O_2 (ppm)	Bati capacity (ml)	Catch No. of Bati/24hr	Composition of fish seed	
								Major carps %	Minor carps, weed fishes %
Barauni	—	5.2	38.0	7.4	6.8	100	1—20	26—40	60—74
Hatidah	2.72	9.1	31.0	7.1	6.8	100	8—115	60–75	24—40
Monghyr	1.65	7.8	33.0	7.2	6.7	100	3—40	40—70	30—60
Sultanganj	3.89	6.5	29.5	6.8	6.6	100	2—30	40—70	30—60
Akbarnagar (Tentis)	0.8	8.4	31.0	6.4	6.0	125	1—35	30–60	40—70
Sonbarsa	3.20	—	30.0	6.8	6.4	125	10—124	70—80	20—30
Bhagalpur	0.70	—	31.0	7.0	6.3	100	4—80	40—60	40—60
Kahalgaon	—	7.0	30.0	6.7	7.4	125	3—70	40—55	45—60
Sahebganj	—	7.0	30.0	7.0	6.0	75	4—90	40—60	40—60
Raj Mahal	2.81	9.0	30.0	7.1	7.0	75	5—100	60—80	20—40
Manikchak	1.46	10.0	29.0	6.8	8.0	75—100	9—115	60—70	30—40
Farakka	—	22.0	29.0	6.9	9.5	100—125	6—90	60—70	30—40

From : Bilgrami & Datta Munshi, 1979.

through a pipe. The mixing zone is about 2 km east from Rajendra Bridge. The effluent contains mainly hydrocarbon compounds, sewage of the township and waste water. Smell of hydrocarbon compounds is evident in this region and leads to both air as well as water pollution. The raw effluents or mixed effluents are black in colour due to thick layer of oil and has distinct odour. The pH was more than 7.0 due to presence of more basic radicals in water. Profuse growth of *Oscillatoria* and *Spirulina* sp., was observed in the contaminated zones. The number of coliform bacteria varied from 27-94/100 ml of the effluent sample. High rate of fish and tortoise mortality was also observed due to these effluents.

Barauni thermal power plant expels effluents in a pond. During rainy season these get mixed with the river.

Bata Shoe Factory is mainly concerned with the tanning of sole and cover leather of the shoe. Effluents from the factory emit nauseating and obnoxious smell resulting in air pollution of a very wide area. On an average 11,000 gallons of unprocessed effluents are discharged per day from the factory, through a channel (3 ft. deep × 5 ft. wide). Effluents contain residues of vegetable and chemical tanning processes and also contain very small fragments of leather which choke the gills of fishes. High rate of fish mortality was recorded in these effluents. They were able to survive only for 48 hours in the mixing zone of effluents with the river. The effluents also had toxic and deleterious effects on the bacterial population. The number of coliform bacteria was recorded to be only 6, 8 and 4/100 ml of effluents sample during May, June and July respectively. In winters the number varied from 23-31/100 ml of water sample. Besides effluents, the factory also expels raw hyde in tons.

McDowell Distillery with a daily production of approx. 13,000 litres of wine remains in operation from December to May. Nearly 2,00,000 litres of unprocessed effluents are expelled out from the distillery every day and channelized into the river near the western side of Rajendra Bridge. The effluent, i.e., collective waste from all units mixed with water from condensers, is red-brown in colour. Its temperature ranges from 60° to 62° and is frothy in appearance. Water temperature in the mixing zone is about 37°C. This site is a source of thermal pollution. The effluents contain waste molasses, nitrogen compounds and barley fragments. High rate of fish mortality was recorded in these effluents, so much so that the fishes died within five hours, on coming in contact with the effluents. The coliform bacteria number varied from 8-84/100 ml of the effluent, indicating its toxicity.

At the point of out-fall of Bata Shoe Factory and McDowell Distillery, only few species of diatoms like *Nitzschia*, *Gyrosigma* and *Synedra* were recorded. Amongst the Zooplanktons, copepods were totally absent from Barauni and Mokamah sites, the region of confluence of industrial effluents with river water. Few forms of *Centropyxis* sp., *Difflugia* and one or two species of rotifers, viz., *Lecane* sp., *Monostyla* sp., and *Keratella* sp., were collected from these spots.

Nearly six kilometres down the stream from the discharge point of effluents of Bata Shoe Factory and McDowell Distillery, the water is not useful for drinking purposes. High rate of fish mortality was recorded. Some dead tortoises were also collected. This is obviously due to toxic effects of effluents near this industrial complex. Mokamah-Barauni industrial complex is a real threat to local aquatic flora/fauna and this problem needs indepth studies.

4.4.2.4.2. Impact of domestic wastes and various of other human activities : Sewage from all the towns along the river is discharged directly into the river without being processed. It was noted that the town sewage with faecal matter supports profuse growth of coliform bacteria. The frequency varied from 175-426/100 ml and 212-542/100 ml of water sample at Monghyr and Bhagalpur respectively. At the above mentioned spots, algal growth was very luxurient which was mainly dominated by the members of Myxophyceae, diatoms, Euglenoids and desmids.

Increasing human population is also one of the major factors in polluting the water of the Ganga. Extensive navigation, transport of boulders specially at Farakka, Sultanganj and Monghyr, bathing, burning of dead bodies, and cattle bathing cause considerable damage to flora and a fauna of the river.

During religious festivals, the number of pilgrims taking dip in the river increases substantially at all the bathing ghats (daily 300-500 people take their bath on non-festival days). Concentration of bathers is very high at Simaria and Sultanganj. Number of washermen (on an average 100) using different types of soaps or detergents, has also increased substantially in the recent past and this greatly retards the water quality at different centres. This problem needs attention as in many cases the points of water supply of the towns is not very far from these polluted areas.

About 120 dead bodies are cremated daily at different ghats at Bhagalpur, Sultanganj, Monghyr and Simaria. There is urgent need to discourage and prevent these activities in order to retain the physico-chemical, biological as well as aesthetic quality of the river.

Data on fish landings (Table 4.16) reflects that fishery of Northern India has been greatly affected after the construction

TABLE 4.16 **Fish landings (in metric tons) during the period 1976-79 at two centres Rajmahal and Farakka, showing the effect of barrage**

	Rajmahal (25°4′N & 87°50′E)						
Species	1976-77	%	1977-78	%	1978-79	%	Annual average
Hilisa ilisha	1.2	1.5	1.9	2.2	2.1	2.1	1.7
Major carps	31.3	38.0	32.4	32.7	43.4	42.9	35.7
Siluroids	42.9	52.0	41.5	48.8	47.5	47.1	43.9
Miscellaneous	6.8	8.5	9.5	11.3	8.6	7.9	8.3
Total landing	82.2	—	85.3	—	101.6	—	89.6
Farakka (24°49′N & 87°56′E)							
Hilisa ilisha	121.5	28	136.8	29.5	148.6	30	135.6
Major carps	83.9	19	98.8	21.5	90.8	19	91.1
Siluroids	180.7	42	196.6	42.8	200.5	41	192.6
Miscellaneous	47.2	11	28.2	6.2	50.7	10	42.3
Total landing	433.3	—	460.4	—	490.6	—	461.6

From : Bilgrami & Datta Munshi, 1979.

of Farakka barrage. This is possibly because adequate arrangements for fish ladder/fish passage have not been made, with the result that movement of popular marine fish, *Hilsa ilisha* to foothills of Himalayas where it goes for breeding is prevented (Bilgrami and Datta Munshi, 1979). Moreover, the barrage has become an important point for indiscriminate fishing. More than a thousand fishing boats are sometimes involved in capture operation of *Hilsa* at Farakka. Since the barrage acts as a barrier against their migration, the fishes get captured before they can breed. As a result *Hilsa* has become a threatened species at present. Besides, there are also some aquatic lives such as dolphin, crocodile, etc., which are reportedly becoming extinct gradually.

CHAPTER V

Lower Ganga Plain

5.1. LOCATION

THE Lower Ganga Plain, in real sense of the term, includes the Kishanganj tahsil of Purnea district (Bihar), whole of the West Bengal State (excluding the Purulia district and the mountainous parts of Darjeeling district) and most of Bangla Desh as well. Though the whole of this Plain is perceived as deltaic, the real delta constitutes about two-thirds of this Plain lying to the south of the Rajmahal—Garo alignment and which together is probably the largest delta in the world (Spate and Learmonth, 1967). The region 21°25′—26°50′N and 86°30′—89°58′ E) embraces an area of about 64,468 km^2, extending from the foot of the Darjeeling Himalayas in the north to the Bay of Bengal in the south (maximum stretch about 580 km) and from the edge of the Chhotanagpur Highlands in the west to the border of Bangla Desh and Assam in the east (maximum stretch about 200 km). In the west, 150 m contour roughly corresponds with the Bihar State boundary while in the east the international boundary with Bangla Desh is maintained (Fig. 5.1). Old mud, new mud and marshes correctly define its topography in a nutshell. This deltaic setting, coupled with perhumid climate, is responsible for the crystallization of a way of life of the people distinct from the rest of the Great Plains.

5.2. THE PHYSICAL SETTING

5.2.1. Geology

Although the making of the Lower Ganga Plain in general

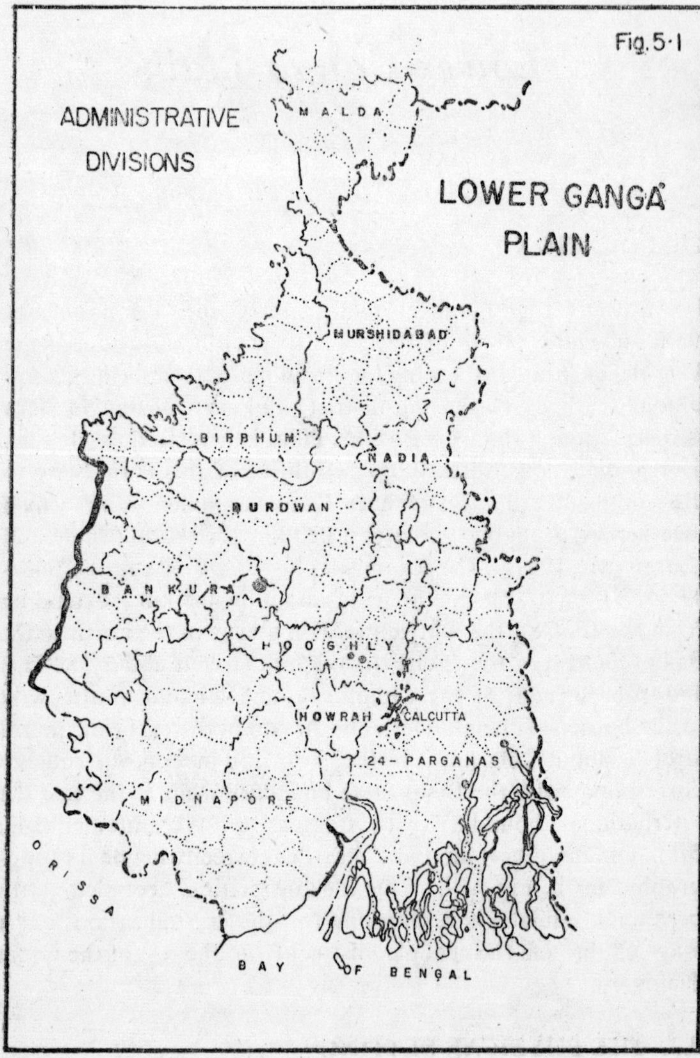
Fig. 5·1
ADMINISTRATIVE
DIVISIONS
LOWER GANGA
PLAIN
MALDA
MURSHIDABAD
BIRBHUM
NADIA
BURDWAN
BANKURA
HOOGHLY
HOWRAH
CALCUTTA
24-PARGANAS
MIDNAPORE
ORISSA
BAY OF BENGAL

and the delta in particular is a product of fluvial action (alluvial and diluvial), the platform for this action was set through the tectonic activities associated with the recent (Tertiary) phase of the earth's organic epoch resulting in the formation of the Trans-Eurasian lofty mountain chains. The formation of the Himalayas created a trench-like trough onto its south (the Sindhu-Ganga trough), while the upwarp of its counterpart—the Arakan Yoma-Andaman-Sumatra arc aided the formation of the Bay of Bengal due to subsidence. Thus, under the two tensional forces, i.e., from the Sindhu-Ganga trough in the north and the downwarping of the Bay of Bengal in the south, the pre-existing narrow neck of the Peninsular landmass foundered, resulting in the formation of the Rajmahal-Garo gap ultimately diverting the waters of the Siwalik river southward into the Bay of Bengal. No doubt, the action was, in part, supplemented by the upwarp of the Delhi Ridge. The existing lateritic tract between the Rajmahal and the Garo Hills may also explain this episode which was followed by the laying down of the recent (late Pleistocene) alluvium and successive southward march of the delta plain. There is conspicuous absence of older alluvium in the delta region, which is quite extensive in the northern parts of the region. This accident did aggravate the actions of the tributaries coming from the Himalayas, and the physical landscape was reshaped to some extent owing to the instability introduced in the river regimes.

This rather major event was sufficient to disturb the isostatic equilibrium of the earth surface locally and the result was the recurrent subsidence, alluviations, river diversions, etc., and shaping and reshaping of the deltaic and other consequent landscapes of the Lower Ganga Plain in general and of the southern part in particular. The well developed beds with erect roots of Sundri trees, peat etc., occurring at about 6-11 m and 150 m below the sea level to testify the regular subsidence of the deltaic region. The western margin is, however, covered by laterites and Gondwana formations, the latter being exposed in the Raniganj of the Damodar basin.

5.2.2. Relief and drainage

The coastal landform (Midnapur coast) reveals the effects

of the changing sea level and the north-easterly winds sweeping the coast. The old coast-line lies about 11 km inland and the Digha beach at present is flat and has a straight shore line which, under the influence of littoral drift, is being eroded away rapidly (Cliff erosion). As many as four parallel lines of dunes can be observed here, some of which rise to about 13 m above the ground and are 45-75 m wide (Niyogi and Chakravarti, 1967). These dunes are partly stabilized. To the east of this shore line lies the deltaic bulge of the Ganga (Sunderbans) where the depositional activity of the stream is prominent and new surface is being continuously added. The easterly shift of the Ganga in 1787 has rendered the area, comprising the district of Nadia and Murshidabad east of the Bhagirathi-Hooghly, a land of dead and decaying rivers. Only in 24-Parganas the delta building is still active.

Of considerable interest is the low altitudinal position of the region where the great agricultural and industrial complex seems to have grown paradoxically. A noted traveller of the 18th century made the following observation for Calcutta "... no worse place could be found in the whole of Lower Ganga Plain for the location and inception of a city of the stature of Calcutta, which became the core of industrial conurbation of the Indian sub-continent" (Kar, 1968). Even a 7 m high tide is sufficient to submerge the region up to Calcutta (inclusive). The saline water expanses in the vicinity of Culcutta testify the recent withdrawal of the sea and occasional march of the tides in this region. Within this low-lying area the gradient is rather imperceptible.

The general slope of the Lower Ganga Plain, as revealed by the drainage system, is to the southeast which is natural in a hinge zone trending NE-SW (Fig. 5.2). The Delta proper has a gradient of less than 2 cm per km, while the areas in its immediate vicinity show gradients up to 40 cm. Only a narrow strip along the northern border of the region exhibits gradients up to 100 cm. The histogenesis of the landform as also largely of the human occupance of this region is a function of the tectonics, changing river channels, sea level changes, etc.

The micro order diversities within the apparently homogeneous physical entity bring out the following physiographic subdivisions.

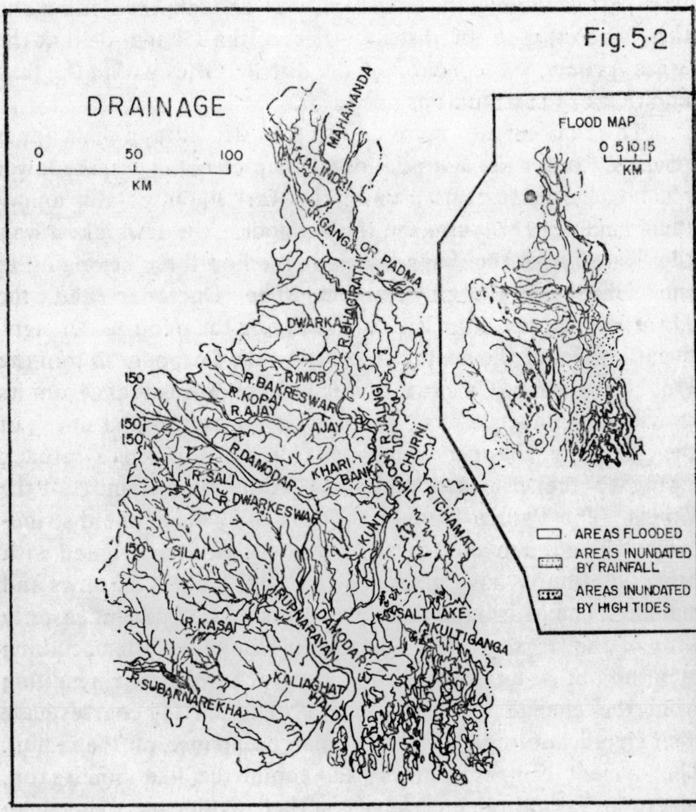

Fig. 5·2

(i) the Northern Plain comprising two distinct land forms (a) the Duars, and (b) the Barind tract; (ii) the Delta proper comprising (a) the land of dead decaying rivers (Moribund delta) in the north (Murshidabad and Nadia), (b) the Active delta of the Sunderbans and (c) the Mature delta (parts of Birbhum, Burdwan, Midnapur and entire districts of Hooghly and Howrah); (iii) the western margin of the delta called the 'Rarh' Plain where the lateritic alluvial landscape along with the coastal scenery at Digha beach has developed.

Drainage system of the region is constituted by the tributaries and distributaries of the Ganga and the Brahmaputra, alongwith some insignificant systems (the Kosi, the Subarnarekha, etc.) discharging into the Bay of Bengal and draining the south-west part of the region. Conspicuously enough, the Padma and the Bhagirathi, both distributaries of the Ganga, delimit the areas prone to the action of the distributaries within the fork and that of the tributaries outside it.

To some extent, more adjusted to structure in their upper reaches, and with aggradational characteristics in the lower reaches, are the streams draining the Rarh Plain, notable among them being the Mayurakshi, the Damodar, the Dwarakeshwar, the Kasai and the Subarnarekha, the first-three belonging to the Bhagirathi-Hooghly system. The Damodar and the Dwarkeshwar are more active in swinging their courses. A significant impact of the southerly sweep of the Damodar to join the Hooghly south of Calcutta, about 150 km downstream from its previous confluence, has resulted in the accelerated silting of the Hooghly channel which had already suffered a setback owing to the eastward migration of the main channel of the Ganga. The Padma-Hooghly fork is the belt of the distributaries where the to and fro swing of the streams associated with other conditions have given rise to 'bils', 'baids', ox-bows and marshes. The Moribund delta also owes to the frequent easterly shift of the streams. The Bhagirathi-Hooghly system, taking off at an obtuse angle, probably occupies a weak zone, resulting from the change in the basin axis. The shifting courses have their significant impact on the human occupance of the region. The ancient centres of trade and commerce like Sonargaon, Satgaon, Tamralipti, etc., have lost their significance, and there is constant threat to Calcutta port and the city as the siltation of the Hooghly keeps increasing.

5.2.3. Climate

The region experiences a hot and humid monsoonal climate. The proximity to the Bay of Bengal in the south, the alignment of the Himalayas in the north and that of the Meghalaya Plateau in the northeast determine largely the climatic character. Irrespective of the general vagaries and mechanism of the monsoon, the spatial and seasonal distributions of the elements such as temperature, rainfall and relative humidity are too uneven.

The average relative humidity is generally high (over 50%).

December and January, show the minimum RH in the coastal tract (Sagar Island 68.5 and 69.5% respectively). During July, August and September the average RH remains over 80% which spatially decreases westward. Mornings are invariably damper with regional variations. The diurnal range is maximum in the month of February (>30%) and minimum in August (<7%) when the monsoon is at its climax.

The rainfall (120 to more than 400 cm) is fairly widespread in the region though with uneven seasonal and spatial distribution characteristic of monsoonal conditions. Out of the four sources of rainfall, i.e. (i) westerly disturbances of winter, (ii) convectional overturning of air resulting in local depression ('Kal Baisakhi') during March-May causing pre-monsoon showers, (iii) Cyclonic disturbances of the monsoon and post-monsoon periods, and (iv) the monsoon currents occurring along the convergence line of the sea-level monsoonal troughs. The last two account for the major precipitation received in the region. But for the southern belt, i.e., the active delta, where August is the wettest month (41.0 cm), July emerges as the wettest month of the region with rains varying between 28 cm in the Bhagirathi-Padma fork and 77.6 cm in the 'Duars'. December is driest month though with relatively damp atmosphere (RH more than 55%).

Spatially the northern and the southern parts, owing to the proximity of the Himalayas and the Bay respectively, experience relatively more annual rains than the central part. It is the area bordering the Chhotanagpur Highlands that experiences minimum rainfall. The erratic nature of precipitation is evident by the fact that even the wettest months receive sometimes rain below 20 cm, similarly the drier months sometimes record exceptionally heavy downpours.

The overall impact of the climatic elements can be interpreted in terms of water surplus and deficit in the region, which have direct correlations with the agricultural economy. The seasonal surplus period, often between mid July and end of October, is a common feature all over the area. There is a general water deficiency period from January to mid-June.

The dry summer is characterised by three air streams, i.e., the northwesterly current from the west, a shallow southerly stream from the Bay of Bengal and a feeble easterly to north-easterly current through the Assam valley, resulting in a marked instability in the atmosphere accompanied by thunder storms, dust storms, squalls, etc.

The winter season is characterised by the sweep of northerly or northwesterly winds. The weather changes are associated with the occasional western disturbances causing some rainfall, but cold waves are rare. Low average temperature (17-24°C) increasing southward and moderate relative humidity (50%) mark the characteristic of the season.

Three micro regions may be distinguished :

1. Perhumid north with over 200 cm annual rain, diurnal temperature range greater than the annual (Jalpaiguri).
2. Humid maritime which includes the coastal parts of Midnapur and 24-Parganas, Calcutta, Howrah and part of Hooghly, experiences rainfall 200 cm and temperature range of 10°C (Sagar Islands).
3. Humid subcontinental includes the rest of the region. There is marked decrease in rainfall (below 150 cm) and temperature range is also high. There is however, more continental effect in the west and two subregions can be defined with representative stations as Asansol and Krishnanagar.

5.2.4. Soils

The soils of Bengal have been studied widely by different scholars. The studies carried out so far lead to the convenient grouping of soils as (i) Laterites, (ii) Red Earths, (iii) Tarai soils, (iv) Alluvial soils, and (v) the Coastal soils (Fig. 5.3).

The lateritic soils (5,888 km^2) are found in the undulating well-drained tract along the Chhotanagpur highlands covering the western part of the region. At places honeycomb structures of the rocks are exposed. Acidic in character (pH 5.5-6.5) and deficient in organic matter these are poorly aggregated and possess low water-holding capacity; on which usually Sal forests thrive.

The transported laterites deposited on the eastern flanks of the lateritic stretch are known as the Red soil (lateritic

Fig. 5·3

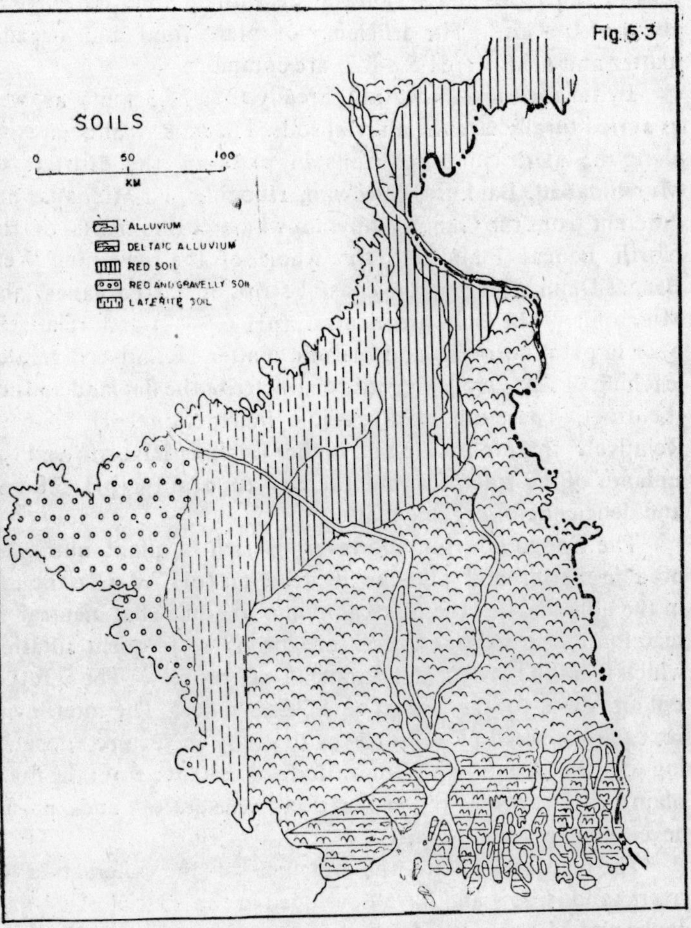

alluvium) and are found in the eastern margins of the 'Rarh' Plain and the Barind tract of Malda and W. Dinajpur covering about 4,963.6 km^2 area. Morum and feldspar and sometimes lime concretions are also observed in the bed. The shallow and coarse-textured soils are acidic in character (pH 6-6 6) and are poor in organic matter and plant nutrients especially nitrogen. Mostly they have been brought under cultivation after deforestation which has accelerated the process of erosion.

The unassorted materials deposited at the foot of the Darjeeling Himalayas are responsible for typical Tarai ('Duar') soils in Jalpaiguri and Siliguri tahsils with an areal coverage of about 66,000 km^2. The deficiency of plant food and organic matter and acidity (pH 5.8-6.7) are common.

By far the most important, areally (28,921.3 km^2) as well as agriculturally, are the alluvial soils. The narrow alluvial strip along the lateritic and red soils in parts of the districts of Murshidabad, Bankura, Burdwan, Hooghly and Midnapur are different from the Ganga Alluvium which covers parts of the North Bengal Plain and the whole of the remaining West Bengal Delta excluding the coastal strips in 24-Parganas and Midnapur. The soils are neutral (pH 6.5-7.2) and relatively poor in plant nutrients and organic matter. Relatively greater leaching of clay, mottling, etc., characterise the flat land soils of the tract. These are mildly acidic in reaction (pH 5.8-6.8). Relatively mature profile and higher leaching have affected the uplands of the tract, leading thereby to acidity (pH 5.8-6 9) and deficiency of organic matter.

The Ganga alluvium however, is rich in plant nutrients, organic matter and alkaline in reaction (pH 7.0-8.2), though in the uplands, i.e., the older alluvium in somewhat neutral in reaction. The riverine tracts are prone to frequent siltation which mars the proper development of profiles. The greyish colour owes to the existence of fine sands. The interfluvial zones are covered by soils, clayey to sandy in texture, depending on the location. With altitude or distance from the flood plain, the process of concretion accelerates and profile developments are also fair.

The coastal soils are the outcome of the interaction of rivers and tides and have developed in the district of 24-Parganas and Midnapur. The soils are saline and alkaline and contain deposits rich in Ca, Mg and partially decomposed organic matter.

5.2.5. Ground water

The region is endowed with ample ground water resources but their occurrence varies seasonally as well as spatially. The porous sandy soils of the riverain areas and the 'Duars' allow the water table to go as low as 30 m during summers (Sen, 1970), especially in northern tract, while in the coastal tract, the lower Hooghly-Rupnarain basins and the Barind tract, it remains within 7 m from the surface. The depth also increases from east to west (>4m-<13m) in the south of the Padma, whereas, it increases in the north.

During rainy season, water table rises up to the surface in most of the areas, notably the Sunderban region, the Lower Hooghly-Rupnarain basin and the Barind tract. Elsewhere also, it remains within 3 m. The condition of water table has led to the prevalence of surface and gravity irrigation-cum-surface types of water management.

There is possibility of richer sources of confined aquifers which may be tapped through deep bore holes. With precise assessment and utilization of such waters seasonal water deficiencies can be made up.

5.2.6. Minerals

Mineral deposits are concurrent with the older formations of the 'Duars' in the north and 'Rarh' in the west, the latter being more important. Raniganj coal fields of national importance lie within this region. The adjoining ore deposits of iron, copper, lead, zinc etc., and their early exploitation, owing to the relative accessibility with respect to Calcutta, led to the initiation of large scale mineral-based industries around Asansol near which a more sophisticated Durgapur complex has been developed. It is notable that the region produced minerals worth about 500 million rupees in 1966, accounting for about 20% of the national mineral production.

Coal is by far the most important mineral as source of power and raw material. The region is credited as the birth place of Indian coal industry (near Sitarampur, 1820) and the Raniganj coal fields are still the second largest in the country. The formations are confined to the Gondawanas lying within the districts of Birbhum, Burdwan and Bankura and continue even beyond into the Chhotanagpur Highlands. The coal seams have been classified into (i) the low volatile Barakar (lower) series, and (ii) the high volatile Raniganj (upper) series. The high grade Raniganj coal is bes tsuited for metallurgical purposes, locomotives and cement works. The copper mines use Raniganj coal as pulverised fuel.

The estimated reserves of coking coal are 288 million tonnes up to 300 m depth and about 548 million tonnes up to 600 m depth (Roy as stated in Sen, 1970). The non-coking coal of superior quality is estimated at nearly 2,759 and 4,838 million tonnes for the corresponding depths. In addition, there is a huge reserve of inferior quality coal in the region.

Iron ores occur in four series :

(1) Banded haematite quartzites containing good quality ore in Buxa (the 'Duars'), (2) Clay limestone of iron stone shales, (3) Lateritic ores of the eastern Raniganj coal fields and (4) Magnetite patches of Archaean metamorphic rocks. Traces of iron smelting works of Agarias in the early historic period are found in Raniganj coal field region and prior to the knowledge of the Singhbhum deposits of Bihar. Raniganj iron ores were considered as the main source for large scale smelting plants in the country. But now, these lateritic deposits are used mainly for building and road-making purposes.

Other minerals of importance are copper near Buxa in the 'Duars', plastic fire clay in Barakar Measures, mines near Gourangdi, Barakar, Suri etc.; sulphide minerals, wolfram, ilmenite, iron, manganese, mica, etc. are scattered in Bankura and Midnapur districts. Dolomite and limestone are found in the 'Duars'. Radioactive minerals occur in Madhabpur and Tilaboni areas of Bankura district. Along the border of the Purulia district, are found graphite and sillimanite. There are also expectations of mineral oil in the Tertiaries underlying the alluvial cover.

5.2.7. Natural vegetation

Natural vegetation of the Lower Ganga Plain can be grouped into several classes according to distribution, growth and expanse (about 9,00,000 hectares). The mangrove and tidal forests in the Sundarbans and humid tropical forests in the extreme north of the region are the only preserves of natural vegetation while the western fringe is covered by tropical deciduous forests mostly in the induced form.

The tropical evergreen forests (1,67,000 hectares) are concentrated in 'Duar' region. Deciduous and scrub vegetation covers greater part (2,99,000 hectares) of the western regions of the Lower Ganga Plain in Midnapur, Bankura, Burdwan and Birbhum. Scattered and isolated patches are also visible in Howrah and Hooghly districts of the Delta Proper. In the tropical evergreen forest the natural vegetation comprises of 'gurjun', the tropical moist deciduous forests comprise of 'Sal', and tropical moist forest includes casuarina and mangrove in this region covering about 4,00,000 hectares (Singh, 1971).

5.2.8. Physical resource base

The Lower Ganga Plain is endowed with vast stretch of flat lands of fertile soil whose fertility is often revived through frequent siltations and which naturally becomes the most potent resources of the region. The soils in association with climate and surface features have contributed to the development of an agricultural landscape dominated by rice-jute culture. The 'bils' and ox-bows, numerous in this humid region, resulting from the frequent changes in the stream course, provide water-fronts for settlement as well as one of the chief sources of fishes which form common diet of the people.. The physiographic diversity from sea to snow is itself an attractive resource for tourists spreading from Digha beach to Darjeeling. The perennial rivers, apart from supplementing the diet of the people through fish catches are also potential resources for irrigation. The tides also facilitate the movement of water-borne traffic since ancient times. The proximity to the Bay provides maritime connection with the outside world. In addition, the Bay is also important for marine fishery.

The 'sundari' trees (Sundarbans) provide base for the indigenous woodworks as well as large scale forest-based industries including paper mills. In addition to enhancing the scenic view of the natural landscape, they provide ground for the presentation of wild life in the region. The huge reserves of basic industrial minerals, i.e., coal and iron, form significant resource base which has stimulated the development of large scale industries amidst vast agricultural lands.

5.3. THE CULTURAL SETTING

In this region of broad homogenity, cultural elements, i.e., population, economy, transport systems, etc , are paradoxically very complex. Densely populated Hooghly basin and the sparsely populated 'Duars', almost wholly rural district of Malda (95.22%) and totally urban Calcutta (100%), immensely agrarian economy of North Bengal districts and the highly non-agricultural (industrial-commercial) economy of the Hooghly side, all stand juxtaposed in the region and make its position much more distinct among all such meso regions in the country.

5 3.1. Population

The region accommodating over 45.2 million people in about 64,000 km^2 area is rather densely populated among all the regions of the country (density 704 persons per km^2), but the distributional unevenness is well-marked at the sub-regional level (Table 5.1). Majority of population (about 60%) is concentrated within the Delta proper covering only two-fifths of area.

The density varies from 345 persons per km^2 in the Bankura district to 31,651 in Calcutta which is highly urbanized. Two more districts showed more than 1,000 persons per km^2 viz., Howrah (2,016) and Hughli (1,127).

The region has shown an increase of 22.08% in its population growth and 36.79% growth in urban population for the decennial 1971-81 (Table 5.1). The highest was in the Nadia district (33.86%), which also showed highest growth rate of urban population (52.65%). Calcutta which is totally urbanized was the district with lowest increase of 4.54% in its total population.

Though the human occupance in the region, like other parts, started with rural agglomerations based on agriculture, the region has at present a significant proportion of urban population (29.36%) which has increased rather steadily since 1901 (12%). Among various regions of the country, it ranks high in urban population. This trend owes to the opportunities providing non-agricultural employment in the urban centres

TABLE 5.1 **Distribution of urban population in the districts situated in the Lower Ganga Plain**

District	Area in km² (000's)	Population 1981		Density per km²	Percentage of Urban to total population		Percentage growth rate 1971-81	
		Total	Urban		1971	1981	Total	Urban
Malda	3.7	2,035,009	97,195	545	4.22	4.78	+26.19	+42.88
Murshidabad	5.3	3,702,869	344,379	696	8.43	9.30	+25.67	+38.62
Nadia	3.9	2,977;013	638,165	758	18.80	21.44	+33.86	+52.65
Twenty-four Parganas	14.1	10,726,751	4,172,619	759	35.15	38.90	+26.95	+40.48
Calcutta	0.1	3,291,655	3,291,655	31651	100.0	100.0	+4.54	+4.45
Howrah	1.5	2,957,464	1,337,335	2016	41.93	45.22	+22.35	+31.95
Hugli	3.2	3,549,817	1,048,752	1127	26.47	29.54	+23.60	+37.94
Medinipur	14.1	6,723,860	574,017	478	7.63	8.54	+22.05	+36.62
Bankura	6.9	2,374,205	181,171	345	7.47	7.63	+16.90	+19.40
Barddhaman	7.0	4,808,886	1,425,659	685	22.78	29.65	+22.80	+59.83
Birbhum	4.5	2,094,756	174,527	461	7.03	8.33	+17.95	+39.88
Lower Ganga Plain	64.3	45,242,285	13,285,474	704	25.45	29.36	+22.08	+36.79

From : Appendix I of Paper 2—Rural-Urban Distribution and Appendix III of Paper 3—Workers and Non-workers, Census of India, Series 1, 1981.

which, apart from offering more assured earning, has the added attraction of safety from natural hazards like floods.

Amongst the district, Calcutta ranks first with 100% urban population, followed by Haora (45.22%), 24-Parganas (38.90%), Burddhaman (29.65%), Hugli (29.54%), Nadia (21.44%), etc. The rest of the districts showed less than 10% urban population with the lowest being at Malda (4.78%).

There are two major areas of urban concentration: the industrial and commercial Hooghly side (the Calcutta conurbation) and industrial-cum-mining Asansol-Durgapur complex. The polarisation of urban population has progressed so much that the conurbation alone accommodates about four-fifths of the region's urban population. On the district level, however, the significance of Calcutta is declining due to want of space to accommodate more population. Eventually, a host of new independent satellites have emerged in the adjoining districts around it. Though by 1981, Calcutta has six times more population than that of 1901, its share in the region's urban population has gone down from about 48% (1901) to 24.78% (1981). The spread of the conurbation outside the district limit and emergence of new urban complexes like Asansol-Durgapur and Midnapur-Kharagpur, are mainly contributing to this trend.

Occupational structure mirrors substantially the regional variation in economy. The region has perhaps the highest proportion of non-workers (about 72%) amongst all the regions of the country. Higher population density, as a main contributor and lack of technical and skilled labour as a subsequent sharer, might have resulted in the lower proportion of workers (28.23%) in the region (Table 5 2).

In spite of considerable urbanization, agriculture still contributes substantially to employment (55.13% of the total working force including cultivators 28.56% and agricultural labourers 26.57%). The rest of the working population is contributed by household industrial workers (4.67%) and workers in other services (40.20%). It is interesting to note that though the region's economy is dominated by agriculture in terms of employment, 1971-1981 decennial recorded a drop of 4.49% in its agricultural workers population which has contributed to an increase in the other working classes, viz., household industry

TABLE 5.2 **Distribution of main workers by cultivators, agricultural labourers, household industry and workers in other services in different districts situated in the Lower Ganga Plain**

District	Percentage of main workers to total population		Percentage of main workers							
			Cultivators		Agricultural labourers		Household industry		Other workers	
	1971	1981	1971	1981	1971	1981	1971	1981	1971	1981
Malda	27.09	27.23	45.58	40.05	33.94	34.40	3.00	4.42	17.48	21.13
Murshidabad	25.91	26.72	39.55	35.19	36.08	33.71	6.83	5.83	17.54	25.27
Nadia	24.77	26.42	37.61	31.68	27.59	28.63	5.42	6.14	29.38	33.55
Twenty-four Parganas	25.87	26.02	24.33	21.75	27.65	21.88	2.34	4.62	45.68	51.75
Calcutta	37.02	34.93	0.02	0.27	0.05	0.30	0.66	2.59	99.27	96.84
Howrah	26.57	26.19	12.07	10.35	20.45	14.51	2.94	7.09	64.54	68.05
Hugli	26.99	28.54	24.83	24.09	30.15	27.77	3.10	4.00	41.92	44.14
Medinipur	26.70	29.27	45.98	48.39	34.43	28.08	2.30	4.15	17.29	19.38
Bankura	28.31	29.07	41.62	41.09	39.75	34.34	3.73	5.01	14.90	19.56
Barddhaman	27.93	28.01	24.23	23.28	30.58	30.32	2.11	3.80	43.08	42.60
Birbhum	26.65	28.10	37.04	37.97	42.26	38.38	2.52	3.72	18.18	19.93
Lower Ganga Plain	27.62	28.23	30.26	28.56	29.36	26.57	3.18	4.67	37.20	40.20

From : Appendices I & II of Paper 3—Workers and Non-workers, Census of India, Series 1, 1981.

1.49% and other services—3% (Table 5.2). Hence, the region shows a tendency to become more industrialized and hence more urbanized, which is probably beyond its capacity in terms of resources and economy, and the general environment of the region.

5.3.1.1. Human settlement

Since the historical period, settlements have dominated the cultural landscape. The hamlets of the Munda-speaking Adibasis (Kols, etc.) refer to human occupance of the 'Rarh' country even prior to the eastward march of the Aryans who followed the Ganga. The introduction of rice culture, irrigation and village council of five, alongwith the entire system of agrarian economy and rural settlements are all attributed to the Mundas, a great prehistoric people of the Ganga Valley. Water channels in the Rarh area were probably the most favoured sites. The Dravidians "were the original inhabitants of Ganga". The Aryanisation of the Lower Ganga Plain was possibly completed before sixth century B.C. The fear of attacks by the invaders and wild animals led to the growth of nucleated and closely packed habitations. In addition to small rural hamlets, there developed a few large settlements as centres of administration, pilgrimage, trade, commerce, etc., at suitable sites, mainly along the rivers.

The chief attracting forces for settlements in the region had been the high banks. In addition to the fertile land, the sanctity of the Bhagirathi water, the then main channel of the Ganga, attracted human settlements. The well-known ancient cities of Tamralipti, Karnaswarna and Salagram, besides others like Kashi, Prayag, Gaya, etc., of the Upper and Middle Ganga Plains, originated on its banks. Other important centres are Aramya (Arambag), Bardhamana, (Burdwan), Bhumisresthi (Bhursut), Bishnupur, Mahananda, Nabadwip, Rajnagar, Vetadda (Belore), etc. But the shifting of the main channel to the east resulted in the decline of the importance of the river as habitable place for some decades. Murshidabad, an early medieval town, situated in the heart of this area on the right bank of the Bhagirathi, is a fine example of a decayed town.

Bishnupur and Nabadwip are equally important for their past glory as ancient seats of culture.

The Lower Hooghly basin again became the main focus of human habitation as is evidenced by the well-known 15th and 16th century towns of Ghusuri, Salkia, Belur, Chitpur, Kalighat, etc., which are now parts of the Calcutta-Howrah complex.

Analysis of the evolution of settlements reveals the relative insignificance of the present thickly populated Delta Proper or more specifically the Active Delta. The monopolistic agricultural economy since long could not compete with the Delta, where economic base has undergone significant change through commerce and industry.

5.3.1.2. Urban settlement

Administration, trade and commerce, and industry—all have contributed to the growth and development of urban centres in this region where religious centres are few. The micro-order topographic prominences in this vast stream-infested land have been sometimes ignored at the cost of the transport facilities, originally the rivers. There can be hardly any parallel to the Hooghly side as example of the 'momentum of early start', which since historic past had been the commercial nucleus of not only this region but almost the whole of the Ganga valley. Later, the modern industrial landscape was appended to this basin which was accelerated by the British policy to undermine the handicraft exports, destroying the indigenous art and craftsmanship of the country as a whole and of this region in particular. By the time the railways and modern road-nets could develop, this river frontage had acquired as much potentiality as to compel these arteries to pass through urbanized belt, occupying the levees of the oscillating Hooghly channel, along whatever narrow high belts were available. The urban establishments have also partly helped in confining this river.

Apart from these two concentrations, the urban centres, mainly as the administrative headquarters (district, sub-division or thana), dot the agrarian landscape as 'islands' and represent typical Indian urban landscape. The former well-grown Dhulian-Bedganga tract in Murshidabad is now reviving with

Baharampur (100, 150) as its centre, aided by the Hooghly side concentration which is already encroaching in Nadia district. Kharagpur-Midnapur region forms the only significant concentiation in the SW which, but for the notorious Damodar and Rupnarain, would have developed better along the main line of the S.E. Railway. Shantipur, Nabadwip and Krishnanagar are notable centres which developed on trade and communication routes as river ports and transicentres. Many towns have declined due to unfavourable circumstances of competition, e.g., Kashripai, Ghatal, Ramjibanpur and Chandrakona in the western margin of the Rarh Plain. On the other hand, many towns like Arambagh (Hooghly), Murshidabad, Baruipur (24-Parganas) and Ondal (Burdwan) gained in importance due to recent development of trade, commerce and transportation activities (Fig. 5.4).

The decline of trade through the Bhagirathi in the Lower Ganga Valley due to partition of Bengal has exerted an adverse effect on the economy of the region.

The map (Fig. 5.4) shows a picture of urban regions of various dimensions. Largest area of influence and high positive growth are the main characteristics of the towns and cities situated in the southern region. In the Calcutta conurbation particularly, the town and city influence is superbly overtapped by Calcutta. The sphere of influence of towns in the North Bengal Plain is comparatively very weak due to lesser development of resources and underdeveloped economic infrastructure. New cities like Kharagpur, Durgapur, Kalyani in the Rarh Plain and Delta region reveal the pattern of transmissional stage of economic development from new centres to the intervening transitional zones.

The growth and emergence of the urban centres, in most cases, being the consequence of successive accrual of non-agricultural activities around a pre-existing agrarian village in an unplanned manner, according to convenience, endowed the settlements with a rather complex morphological character with respect to functions as well as structures.

The region now has around ten urban agglomerations of Class I cities of which, four are situated along or in the vicinity of the Hooghly estuary. Population as per 1981 Census, growth rates in the last two decennials, working populations and its

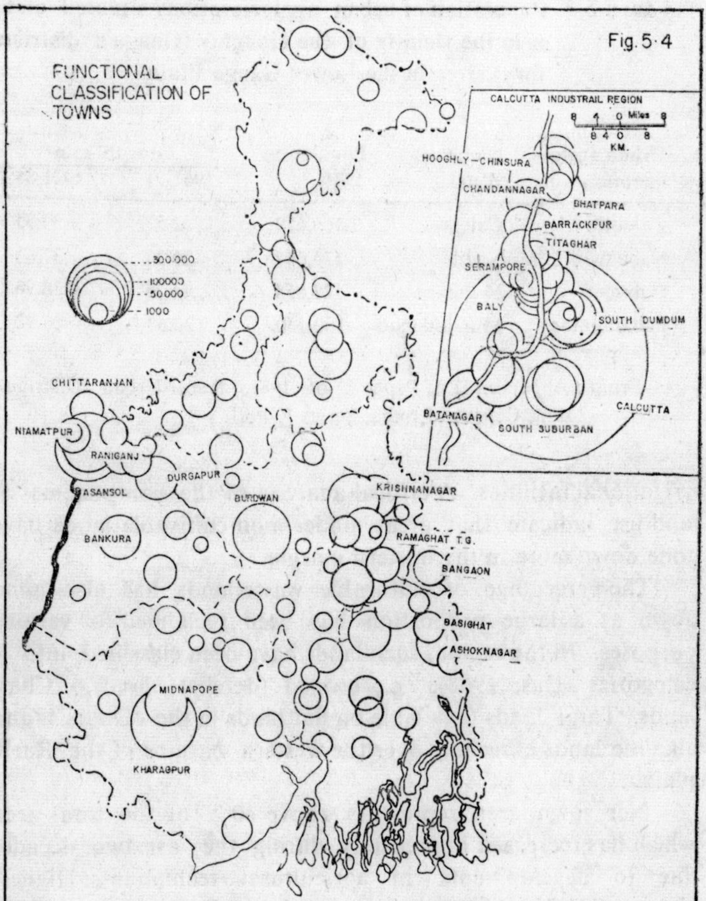

distribution in different categories, for these four cities are given in Tables 5.3 and 5.4.

5.3.2. Agricultural economy

About 55% of the total working population is still engaged in agricultural pursuits. About 70% of the total area of the region is under cultivation. In general, North Bengal Plain is primarily agricultural. Regional variations in agricultural resources are associated with topography, soils, rainfall and

TABLE 5.3 **Population of urban agglomerations situated along or in the vicinity of the Hooghly (Ganga's distributary) river in the Lower Ganga Plain**

Urban agglomeration	Name of district	Population 1981	Growth rate 1961-71	Growth rate 1971-81
Calcutta	Calcutta	9,165,650	+22.57	+30.35
Nabadwip	Nadia	129,647	+29.29	+37.62
Habra	24 Parganas	127,855	+44.25	+36.96
Baharampur	Murshidabad	100,150	+26.63	+26.92

From : Appendix II of Paper 2 of 1981, Rural-Urban Distribution, Census of India, Series 1, 1981.

irrigational facilities. A critical analysis of the components of land use indicate that areas under non-cultivable lands have gone down more in the western margin.

The percentage of culturable waste-lands has also gone down as a large proportion has been reclaimed for various purposes. In the region, such lands have been classified into 4 categories (Das, 1968), i.e., eroded (derelict) lands, 'Char' lands, 'Tarai' lands (available on flat lands in the 'Duars') and alkaline lands especially over the western margins of the 'Rarh' plain.

Net sown area represents above 60% of the total area which has increased appreciably during the last two decades due to developments in agricultural techniques. Highest increase (21-31%) in net sown area has taken place in West Dinajpur and Nadia. In the 'Duar', Jalpaiguri and Cooch-Behar show an increase of 5-7%, while Burdwan and Midnapur in the Delta regions account for 2-3% increase. Some decrease (5-10%) in net sown area has taken place in some delta districts such as Howrah, Hooghly, 24-Parganas and Murshidabad.

About 50,000 hectares of decrease in current fallow has been recorded in Jalpaiguri, Midnapur and Nadia, mainly due to climatic fluctuations, limitations in wetness of soils and other socio-economic factors operating on individual holdings.

Double cropping has been practiced since 1950's. Generally 'aus' lands are utilised for double cropping for growing

TABLE 5.4 **Distribution of population by workers and non-workers in the urban agglomerations situated along the Hooghly (Ganga's distributary) river in the Lower Ganga Plain**

Urban agglomeration	Total main workers	Workers as per cent to total population	Cultivators	agricultural labourers	Household industry	Other workers	Non-workers
Calcutta	2,796,382	30.51	13,518	26,575	1,26,724	2,735,510	6,262,323
Nabadwip	34,161	26.35	617	968	5,225	28,582	94,255
Habra	30,203	23.62	697	1,827	1,616	26,442	97,264
Baharampur	24,761	24.72	274	589	792	23,541	74,954

From : Appendix IV, Paper 3 of 1981—Workers and Non-workers, Census of India, Series 1, 1981.

vegetables and rabi crops. The practice is more popular in the delta districts of Hooghly, Nadia and Murshidabad.

5.3.2.1 Irrigation

Irrigation facilities in the region are inadequate. Though the average rainfall is fairly high, irrigation is essential for agricultural prosperity due to irregular monsoons and dry winters. Moreover, water balance is adverse in the Lower Ganga Plain region. The completion of Damodar Valley Corporation and Mayurkashi Projects have now provided large scale irrigation. The highest percentage (60%) of irrigated area by Govt. canals lies in Burdwan, followed by Birbhum and Bankura (30%), Hooghly (29%) and Murshidabad (14%). Irrigation from ponds, 'bils' and other sources is more popular in Murshidabad, Midnapur, Howrah (40%), followed by Birbhum and Bankura (32%). Well irrigation is not so popular except in Nadia and Malda where 20% of irrigated area is under wells. Double cropped areas are irrigated in the 'Duars' and delta area, e.g., Nadia and Malda (30% each), Hooghly (17%), Murshidabad (15%) and 24-Parganas (11%).

5.3.2.2. Cropping pattern

Rice is by far the most important crop, covering 75% of the net sown area and 85% of the total cropped land. It also accounts for 90% of irrigated area. The yield and irrigated area correlation is well marked (co. coeff. 0.89). 'Aman' variety of rice dominates throughout the region while 'Aus' is significant in Malda, Nadia, Murshidabad and Bankura. 'Boro' is cultivated on hardly 1% of cropped area and is confined to the marshy tracts of Howrah, Hooghly and 24-Parganas.

Jute as the cash crop stands second in importance with 4.3% of cropped area. It maintains, in general, second rank in Active Delta and the Duars. Elsewhere, except along the western margin, pulses rank next to rice. Til in the western margin and vegetables in the suburban zones, are other crops of significance. Among the four types of agricultural lands, 'Sol' and 'Kanali' are devoted to 'Aman', Baid to 'Aus' and

'Til', mustard and groundnut are grown in 'Danga' which is also used as pasture.

5 3.2 3. Crop-association and farming

Cropping in the Lower Ganga Plain is stable due to level land and firm rotational practices. But there is enough scope for reorientation and new adaptations to the environments. Rice has been the most outstanding crop in terms of area and output. Rotation of crops indicates intensity of cropping in an area. It is noticed that two distinct rotations are prevalent in the region, i.e., (i) 'Aman'-fallow—'Aman' system in the lowlying fields, and (ii) 'Aus'-Jute-rabi pulses or vegetables system on the upland fields. In the lowlying areas ('Aman' rice fields), there is little scope for crop rotation. But over 'Aus' fields several rotational practices may be developed, especially where maize, potato, vegetables, jute and rabi pulses are grown, provided irrigation and other facilities are available. Considering the crop components, five crop associations in the region have been determined. Rice is the principal component which has got its privileged position on account of favourable factors for its cultivation. The North Bengal Plain shows two broad groups of association : (i) Rice-Jute, Castor seed, Potato; and (ii) Rice : Jute, Rape and Mustard, Barley, Maize. It is quite apparent that these crop associations display some variation in cropping after the second rank. On the basis of yields of various crops, agricultural landscape has been rated as Fair and Medium. The land use components do reveal significant characteristics of the regional response in cultivation, double cropping and areas under irrigation (Fig. 5.5).

5.3.3. Fisheries

The preponderance of water bodies, natural and artificial, and to some extent other environmental conditions like climate, acute shortage of agricultural land and lack of other resources of animal protein, have led to the adaptation of fisheries in the region. Culture fisheries in tanks and ponds roughly constitute

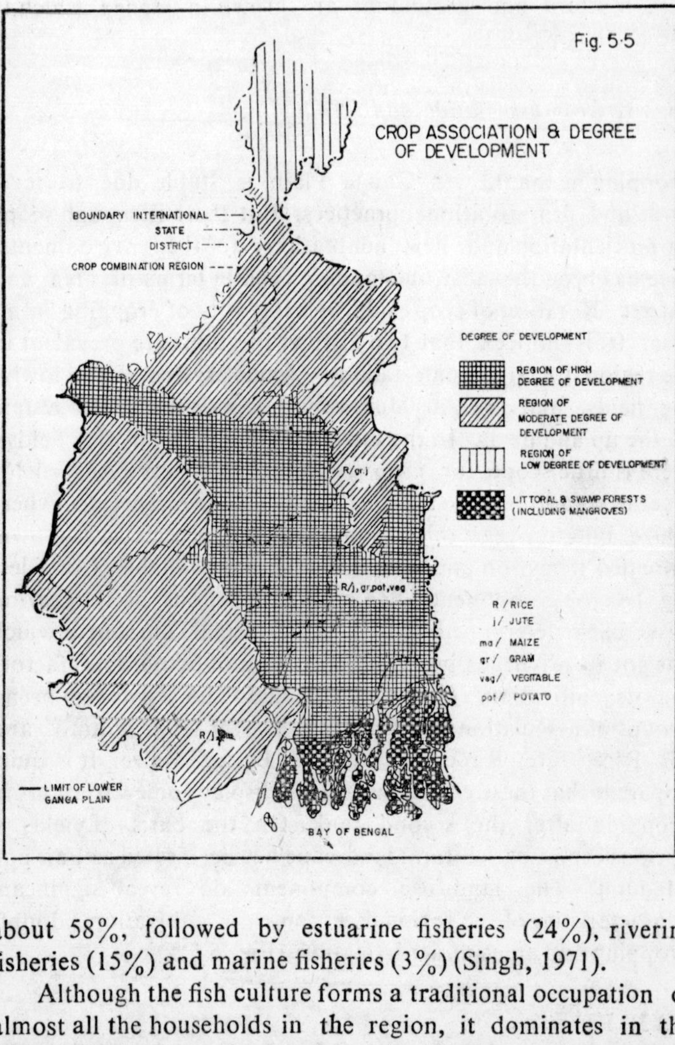

about 58%, followed by estuarine fisheries (24%), riverine fisheries (15%) and marine fisheries (3%) (Singh, 1971).

Although the fish culture forms a traditional occupation of almost all the households in the region, it dominates in the Delta Proper and the southern Barind tract. The common culture fisheries constitute Rohu (*Labeo rohita*), Catla (*Catla catla*), Mrigala (*Cirrhinus mrigala*) and Kalbasu (*Labeo calbasu*). Apart from the fresh water 'bils' and tanks, sewage-fed fisheries in the salt lake areas of Calcutta are notable where the yield per hectare is over 60 maunds. Riverine fisheries are important in 24-Parganas, Murshidabad and Midnapur which account for more than 90% of riverine catch. Hasanabad, Itindaghat, Port Canning, Diamond Harbour, Kolaghat and Lalgola are the main collecting and transport centres. Most of the riverine as well as the culture fishes are consumed locally. 'Hilsa' is the most important commercial fish, common in the Calcutta market. Sunderban areas are the favoured sites for estuarine fishery.

The Sunderban area abounds one of the major estuarine systems in India: The Hooghly-Matlah estuary. The estuary extends from its mouth up to Nabadwip in the north and is bound on the east by portions of rivers Ichamati, Harinbhanga and Gosaba and on the west, by river Rupnarain and certain other streams. The Hooghly or the Bhagirathi which is the main distributary of river Ganga has 5 tributaries, viz., Jalangi, Churni, Damodar, Rupnarain and Haldia. In the lower region of this estuarine system, there is a number of marshy islands between which are present silted-up distributaries appearing as mere intrusions of the sea into the conglomeration of islands. The more prominent distributaries are: the Saptamukhi, the Thakuran or Jamir, Matlah which is confluent with the Bidya, the Gosaba and the Ichamati. The total approximate area of Sunderbans estuarine waters is 2,340 sq km (Mitra, 1967—From Jhingran, 1977).

The Estuarine Division of Central Inland Fisheries Research Institute, Barrackpore has been concentrating on the study of fishery and hydrobiology of the Hooghly-Matlah estuary. For the purpose of estimating fish landings by sampling, the Hooghly-Matlah system has been divided into the following zones:

Zone I—Nabadwip to Konnagar, the upper zone or the fresh water zone.

Zone II—Konnagar to Diamond Harbour and called the middle zone or the 'gradient zone'.

Zone III—Diamond Harbour to the mouth of the estuary including entire lower Sunderbans. This is the lower or marine zone of the system.

Zone IV—River Rupnarain.

Zone V—River Matlah around Port Canning.

Table 5.5 shows fish landings along with the total estimated catches from the Hooghly-Matlah estuary for the years 1964-65 to 1969-70. The most productive area in the entire estuary is Zone III which contributes 70-80% of the total annual landings of the entire estuary.

TABLE 5.5 **Total annual fish landings (tonnes) from the Hooghly-Matlah estuary**

Period	Total landings (tonnes)
March 1964 to Feb 1965	10413457
March 1965 to Feb 1966	6516096
March 1966 to Feb 1967	6766338
March 1967 to Feb 1968	8805474
March 1968 to Feb 1969	8274970
March 1969 to Feb 1970	8258868
March 1970 to Feb 1971	10992827

From : Jhingran, 1977.

From the consumers' viewpoint the most important species of the Hooghly-Matlah estuary are *Hilsa ilisha, Lates calcarifer, Polynemus paradiseus, P. indicus, Mugil parsia, M. tade, Pama pama, Sillaginopsis panijus* and prawns (Gopalakrishnan, 1968).

With a coastline of about 65 km there is ample scope for marine fisheries but due to lack of equipment and adequate craft and berthing facilities, only 32 km of coast between Junput and Digha are being effectively used. The catches comprise 30% prawns, 40% quality fish and 30% low grade fish. Junput is developing as an important fishing centre. There is little impetus for off-shore fishing.

The fish catch from the area does not meet local consumption, hence the region has to import about 30,000 tonnes per annum from other regions including Bangla Desh. To achieve this end, good storage facilities are essential. At present all the inter-regional fish movement services is provided by railways which find it difficult to cope with the increasing demand.

Popularity of fish in diet can very well be emphasized with the per capita annual consumption of more than 4 kg against the national consumption figure of less than 3 kg. With mechanised and organised development of fisheries the region may not only become self sufficient but also fetch foreign exchange by export.

5.3.4. Industrial economy

The region is richly endowed with materials, mainly coal, water and man-power and has access to the mineral belt of India as also to other raw materials. The main industrial concentrations have developed in the Hooghly side and Asansol-Durgapur belts, which together account for over 90% of the total industrial employment of the region. About eleven groups of industries have sprung regionally (Fig. 5.6).

Major textile industries are located in 24-Parganas and Nadia (Calcutta industrial belt), while smaller establishments are in Bankura and Murshidabad. Calcutta industrial belt is responsible for employing about 80% of the total workers in the delta region in the manufacture of jute textiles, the highest portion being in Howrah (79%), in Hooghly (72%). Among all groups of chemical industries, the highest proportion of workers is engaged in industries manufacturing synthetic rubber and rubber products, the spatial break up being 41% in Calcutta, 55% in 24-Parganas and 63% in Howrah (Singh, 1971). The metal based industries are also concentrated, with highest proportion of all categories of factories in Howrah (387), Calcutta (275), 24-Parganas (205) and Burdwan (20). The recent figures however are not easily available for all the groups of industries for comparing the increase that has taken place in the last two decades. Iron and steel industries employ the highest percentage of workers within those in metal-based industries—Burdwan and 24-Parganas (>90%), Calcutta (>80%) and Howrah (>70%). The next important metal-based industries are aluminium and structural steel products.

Other industries such as, wood-based, leather, food-processing, transport equipment, etc., are concentrated in different parts. Among these, the North Bengal Plain accounts for mostly wood-based, beverages and tobacco and food-processing industries.

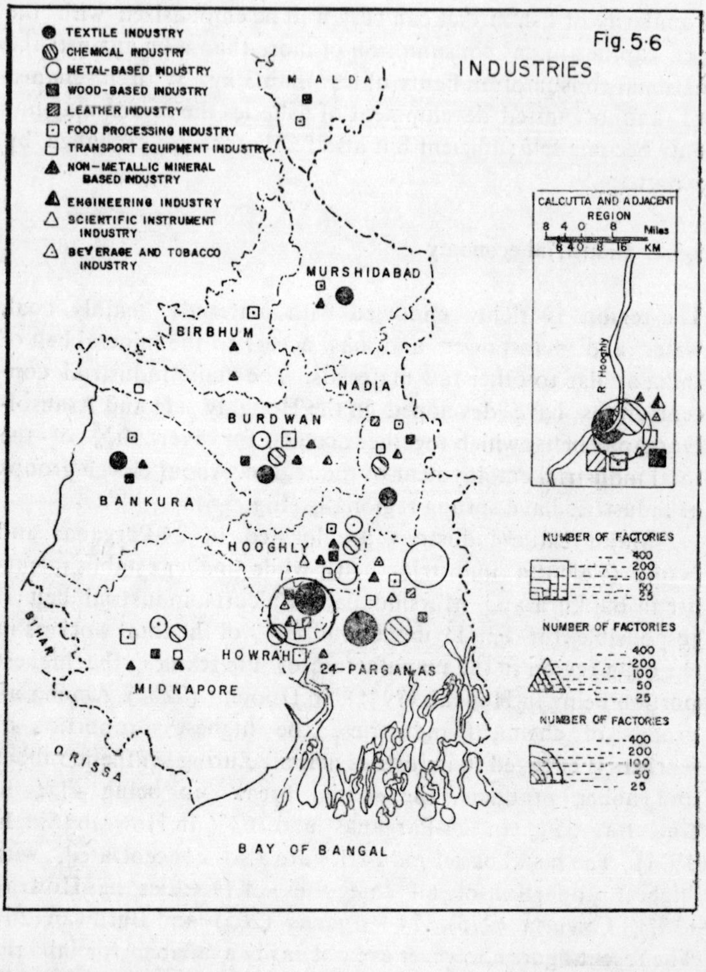

The Hooghly industrial belt (rather the delta proper) is a zone of diversified industries while the Asansol-Durgapur complex is specially noted for metal and chemical-based industries. The regional disparity in industrial landscape is obvious due to the uneven distribution of industrial resource base and under-utilization of economic infrastructure. However, the Delta proper has a wider base for industrial development and location of diversified industries, if considered on the basis of distribution of all classes of industrial establishments and proportion of workers deriving livelihood from them.

5.3.5. Transport and communication

Quite in harmony with the location at the transshipment point of water and surface-borne traffic, a fairly good transport net with wide spatial variation has been developed in the region even against a rather unfavourable terrain (Fig. 5.7). The transport arteries of the region, in addition to catering to the international movement of goods, have to haul huge amount of inter- and intra-regional traffic, viz., cereals, jute, tea, coal and other minerals and industrial products. Prior to the advent of the quicker and more assured surface transport arteries like railways and roads, the perennial streams like the Ganga and the Brahmaputra served as the link between the ports along the Hooghly basin and the hinterland.

The river-side centres experience deterioration due to (i) the oscillatory channel characteristics, (ii) the advent of the more stable surface transports which have the advantage of penetrating the forest-clad interiors, and (iii) the political partition of the region which disrupted the whole of the river transport affecting much of our jute and tea shipment from Assam and the North Bengal Plain. Though there are still 27 partly or wholly navigable streams of different sizes and 10 canals within the nation, the northern water routes excepting the Mahananda, are obstructed by the international border with Bangla Desh.

Though the region has the highest density (0.42 km/km^2) of the rail-roads in the country, 25% of its area still lies at a distance of more than 16 km from any track. These chunks lie in the SE Active Delta, the 'Duars' and the undulating Rarh except the Asansol-Durgapur industrial belt. Surface links with Assam and the Duars have become more circuitous after the partition. For this region, where huge traffic originates and terminates, the existing tracks are not adequate. Calcutta-Howrah is the main convergence of the railways. Electrification, bridging up of the interrupted links and modernisation of the tracks will go a long way to improve the rail transport in the region.

Like the railways the region has also the highest density of roads of all kinds. The 25,000 kilometres of roads (excluding the 1200 km of village roads) include partly or wholly, at least 5 national highways (Nos. 2, 6, 31, 34, 35), including the Grand

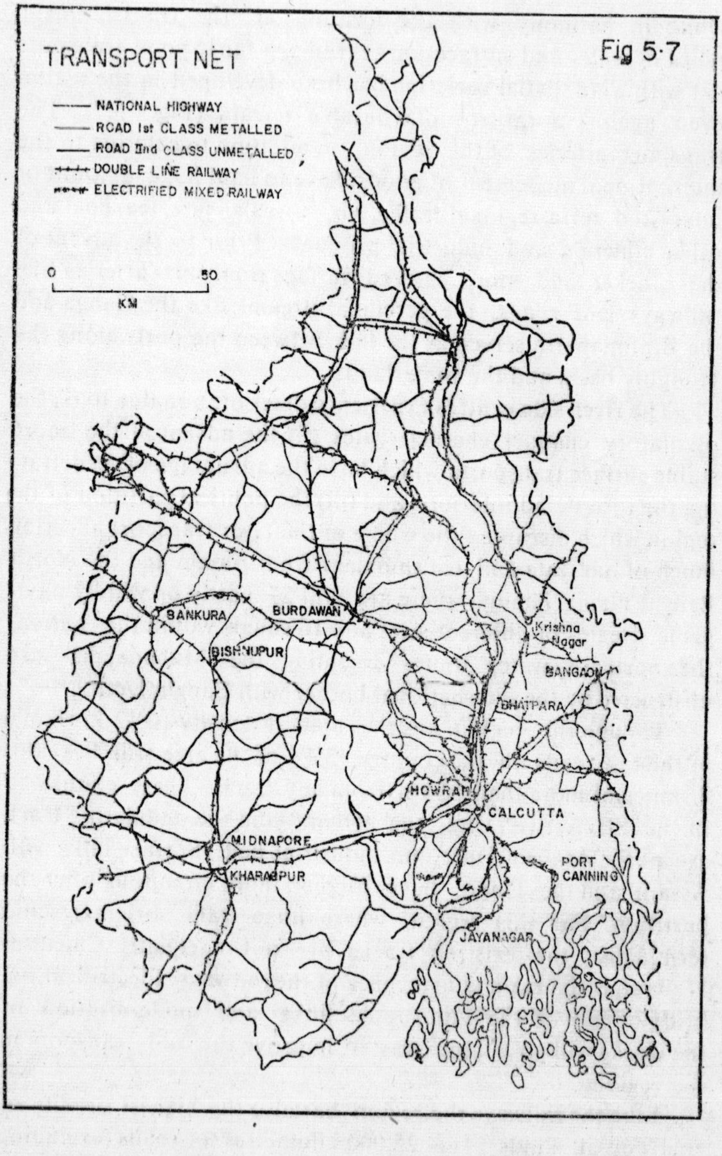
TRANSPORT NET
Fig 5·7
NATIONAL HIGHWAY
ROAD Ist CLASS METALLED
ROAD IInd CLASS UNMETALLED
DOUBLE LINE RAILWAY
ELECTRIFIED MIXED RAILWAY
0
50
KM
BANKURA
BURDAWAN
BISHNUPUR
Krishna Nagar
BANGAON
BHATPARA
HOWRAH
CALCUTTA
MIDNAPORE
KHARAGPUR
PORT CANNING
JAYANAGAR

Trunk Road (NH-2). Swamps and oscillating river channels provide the greatest hindrance to smooth transport.

With Dum Dum as a busy national and international airport, the region enjoys a significant position on the air map. Apart from handling the distant traffic, it has also to provide air connections to Meghalaya, Manipur, Tripura, Nagaland, and Assam. Local flights to the Duars are also available.

5.4. PROBLEMS IN THE REGION

The region endowed with rich alluvial soil, abundant basic raw materials, numerous perennial streams and a seaface attracts a large number of people in search of their livelihood. It is now confronted with certain acute problems, both (i) Physical, and (ii) man-made.

5.4.1. Physical

The Lower Ganga Plain, similar to the preceding regions upstream, also has to face the brunt of floods from the Ganga and the other riverine systems in the region. Flood damages in terms of actual losses of property, human lives, etc., and the cost of damages involved, for the period 1953-81 are given in Tables 5.6 and 5.7. Since the region comprises almost the entire State of West Bengal, the statement of flood damages for the state is referred to as that for the region. It is revealed that 231.8 lakh ha area, 707.8 lakh population, 75.6 lakh ha of cropped land, 4.5 million houses, more than 3 lakh cattle and 5,242 human lives have been damaged or lost in the floods from 1953 to 1981. The cost of total damages amounts to Rs. 955.31 crores, for the same period. It can also be seen from the tables that the damages were negligible or low in the earlier period (1953 to 1966) but there was a sudden rise in the fury of the floods resulting into increasing damages in the later period (1967-1980). The maximum damage was reported in the year 1978. Recurrence of floods causing heavy damage in the recent past can be attributed to the increased human occupance and increased activities as also due to heavy deforestation in the upper stretches in the area, contributing to the silting of the estuarine zone.

TABLE 5.6 **Flood damage during 1953 to 1981 in the State of West Bengal**

Sl. No.	Year	Area affected in lakh ha	Population affected in lakhs	Damage to area in lakh ha (crops)	Damage to No. of houses	Cattle lost	Human lives lost
1.	1953	Neg.	NA	NA	NA	1000	NA
2.	1954	4.2	NA	NA	NA	1500	142
3.	1955	3.5	NA	NA	600	20	2
4.	1956	26.5	9.0	4.0	199556	2004	55
5.	1957	11.6	1.0	Neg.	225	NA	2
6.	1958	4.5	13.9	1.3	42935	5092	11
7.	1959	11.2	4.0	6.1	504464	8473	107
8.	1960	1.9	0.8	0.2	1800	16	NA
9.	1961	2.0	0.1	0.1	NA	NA	NA
10.	1962	1.6	2.5	0.2	2645	280	Nil
11.	1963	0.9	1.0	0.3	1300	Nil	Nil
12.	1964	1.0	1.9	0.2	8298	14	Nil
13.	1965	1.2	7.5	0.6	4457	28	Nil
14.	1966	1.3	3.6	0.3	7271	1	2
15.	1967	9.4	16.3	4.2	84440	745	27
16.	1968	21.7	81.3	9.5	269955	67758	2730

17.	1969	5.3	11.6	1.7	87845	102	30
18.	1970	20.1	97.7	9.0	501170	4014	85
19.	1971	20.6	111.8	10.8	669709	3164	153
20.	1972	4.4	17.2	1.4	75453	2521	19
21.	1973	8.7	29.3	4.2	91879	1291	28
22.	1974	3.8	16.6	2.1	99049	1524	33
23.	1975	2.0	7.1	0.4	12985	28	7
24.	1976	13.0	16.4	1.6	125913	3646	67
25.	1977	15.5	63.1	3.5	88572	372	25
26.	1978	30.8	156.3	11.4	1361338	221826	1596
27.	1979	0.2	1.5	NR	1051	NR	1
28.	1980	3.8	26.2	2.1	268914	1095	65
29.	1981	1.1	9.9	0.4	19897	1	55
Total		231.80	707.8	75.6	4531721	326515	5242
Maximum		30.8 (1978)	156.3 (1978)	11.4 (1978)	1361338 (1978)	221826 (1978)	2730 (1968)
Average		8.0	27.2	3.0	174297	12558	202

NA = Not available, NR = Not reported, Neg. = Negligible.

From : Central Water Commission, New Delhi.

TABLE 5.7 **Flood damage in lakhs of rupees during 1953 to 1981 in the State of West Bengal**

Sl. No.	Year	Crops value	Houses value	Public utilities	Total damage
1.	1953	5.0	1.0	10.0	16.0
2.	1954	760.0	100.0	200.0	1060.0
3.	1955	50.0	NA	100.0	150.0
4.	1956	742.0	NA	33.0	775.0
5.	1957	0.7	NA	NA	0.7
6.	1958	554.9	30.2	NA	585.1
7.	1959	4400.0	722.0	NA	5122.0
8.	1960	NA	NA	NA	NA
9.	1961	NA	NA	NA	NA
10.	1962	61.6	2.3	7.5	71.4
11.	1963	15.0	0.6	0.3	15.9
12.	1964	73.1	11.6	43.6	128.3
13.	1965	284.6	3.9	1.9	290.4
14.	1966	159.0	15.0	2.0	176.0
15.	1967	1962.0	239.0	188.0	2389.0
16.	1968	5355.7	2593.9	657.0	8606.6
17.	1969	7802.3	349.2	288.0	1439.5
18.	1970	6088.6	1125.4	1485.4	8699.4
19.	1971	12393.2	2125.6	1381.3	15900.1
20.	1972	1009.0	192.1	304.6	1505.7
21.	1973	3493.0	334.2	597.5	4424.7
22.	1974	2626.1	15.2	41.4	2682.7
23.	1975	394.1	33.2	24.1	451.4
24.	1976	1690.4	260.4	151.2	2102.0
25.	1977	3947.2	NA	36.3	3983.5
26.	1978	20735.0	9.1	6115.5	26859.6
27.	1979	NR	NR	0.5	0.50
28.	1980	3337.5	2626.7	1837.4	7801.5
29.	1981	214.2	64.7	15.1	294.0
Total		78154.2	10855.3	13521.6	95531.0
Maximum		20735.0 (1978)	2593.9 (1968)	6115.5 (1978)	26859.6 (1978)
Average		3005.9	493.4	563.4	3538.2

NA=Not available, NR=Not reported, Neg.=Negligible.
From: Central Water Commission, New Delhi.

5.4.2. Man-made

The factors which induced rapid economic development in the Hooghly region have themselves generated a chain of complex problem in the very wake of the vigour of development. On the regional level development was so lop-sided and uncontrolled that even in the exceptionally developed areas (Calcutta) the attendant problems could not be resolved. The most affected area of the lot is the Hooghly estuary which has been the life-line of the region for its growth and development so far. Various studies have been carried out along the Hooghly estuary as regards the impact caused by industrialization on the estuarine eco-system as a whole.

A most recent report on the estuary is that of Ray and Laha (1981) who have studied the effects of industrial and municipal wastes on the trend of biological characters and fisheries (data from 1962-63 to 1977-78) in the Hooghly estuary, in a 158 km stretch between Nabadwip and Baj Baj. According to them the actual volume of water presently flowing in the river Hooghly the largest and widest estuary of the Indian subcontinent is alarmingly low. The river receives domestic and industrial wastes round the year and is subjected to varying dilutions during different seasons under different tidal phases. Changes in the hydrography of the estuary and direct and indirect effects of polluted waste-water discharged into the Hooghly are not always congenial for survival, growh and reproduction of biomass. The problem is more acute during dry summer months, owing to the loss of homeostatic power resulting in considerable destruction of fish food organisms. The estuary once provided excellent spawning beds for some of the most priced fishes like hilsa, mullets, prawns, etc. Natural resources of seed of these species are now rare. As a result the fishery in the estuary is adversely affected. Conditions are further complicated in the years of late arrival of monsoon and prevalence of drought conditions.

5.4.2.1. Sources of waste-water and their pollution load

5.4.2.1.1. Industrial wastes : A comprehensive survey (Ghosh *et al.*, 1973; Ghosh and Bagchi—unpublished data) reveals that 96 industries concentrating in the lower zone between Bandel and Baj Baj, discharge their effluents into the estuary. These industries are Jute (55), Paper and Pulp (6), Textiles (5), Tannery, Distillery and Yeast (3), Thermal Power Station (6), Oil (Petroleum—3 and Oil Refinery—1) and miscellaneous (Paints and Varnishes—3, Polythene, Hydrogenated Vegetable Oil and Soap—3, Safety Match, Metal and Steel—2, Chlorine, Rubber Chemicals, Insecticides, Phosphate Fertilizer and Shellac).

Nearly 80% of these industries discharge their wastes directly into the estuary almost in an untreated condition. The average daily flow of waste-water from various industries is 0.358 million m^3, out of which Paper and Pulp Industry alone contributes 49.86%. The average total pollution load per day from all the industries calculated in terms of BOD_5, total solids, suspended solids and dissolved solids are 6.188, 894.053, 412.142 and 481.91 tonnes respectively. Pulp and Paper industry contributes 28.59 t BOD_5 (43.2%), 547.40 t total solids (61.2%), 299.20 t suspended solids (72.5%) and 248.20 t dissolved solids (51.5%). The tonnage for BOD_5, total solids, suspended solids and dissolved solids for miscellaneous wastes are 14.92 t (13.95%), 164.235 t (7.08%), 55.305 t (5.23%) and 108.43 t (8.64%) respectively. The total pollution load has been discussed by Ray (1980) in greater details.

5.4.2.1.2. Domestic and Municipal Wastes : Combination of domestic and municipal wastes in the estuary is significantly high—0.7955 million m^3/day. The pollution loads are 40.08, 1424.36, 645.40 and 778.96 t/day of BOD_5, total solids, suspended solids and dissolved solids respectively.

Combining the total of industrial, domestic and municipal wastes, the volume of effluent disposed and their corresponding loads amount to 1.1535 million m^3/d equivalent to 106.268 t BOD_5. 2318.413 t total solids, 1057.542 t suspended solids and 1260.87 t dissolved solids.

5.4.2.1.3. Bacteriological load : Counts for coliform bacteria in the lower stretches of the Hooghly estuary were found to vary between 8.85 and 201×10^4 (MPN in 100 ml) during monsoon and between 6.24 and 85×10^4 during dry weather flow (NEERI, 1978). The upper stretch around Palta registered values ranging between 0.5 and 0.7×10^4 (FAO, 1967). This shows that the values at the lower stretch are far above the standard limits (should not exceed 5000) and distinctly signify unhealthy ecosystem.

5.4.2.2. Effect of waste-water on the biotic life

5.4.2.2.1. Pollution in general : The Hooghly estuary seems to have high buffering capacity since the fluctuation in pH value of the estuarine water is low in the entire stretch except at the effluent outfall region. During post-monsoon period (November-February) the estuary returns to its normal condition and a sort of stabilization is observed as it receives minimum disturbance, particularly in the upper stretches. From February onwards, until the beginning of June (monsoon), the proportion of freshwater gradually reduces. Large quantities of non-conservative and conservative material disposed along with the effluents, not only affect the fishery but also general biotic life by altering food chain and suitability of river bed for spawning. High temperature (31.7-43.2°C) of effluent water reduces the level of dissolved oxygen by the process of biodegradation of easily decomposable material which remain both in suspension and settled around the outfall. Although the average value of DO studied for six years 1967-72 indicated saturation above 65% on several occasions, the stretch between Nawabganj and Baghbazar (lower stretch) exhibited saturation between 46 and 58%. WHO and Engineering Consortium while working for CMPO in Project E 170, opined that 60% of the community waste-water needs to be treated to keep a minimum DO of 4 ppm (Ray, 1980).

The effect of waste-water on aquatic biomass was observed to occur mainly around the outfall. Cumulative effect of waste-water in the main estuary was noticeable during summer season, when headwater supply reduced considerably leading to low dilution ratio. This resulted in reduction of plankton concentration. Besides, transitional period between high and low tide and vice versa of each tidal phase, particularly during neap tidal phase, the effluents caused heavy damage to planktonic organisms. Table 5.8 shows the percentage reduction of plankton around the outfalls of some important industries.

During winter, plankton density was higher while in summer months pollution impact was felt more intensely. During bore tide the plankton concentration remained almost uniform in the estuary (Ray, 1980).

Seasonal variation in primary productivity (gross and net) observed in the entire stretch of the estuary was found to be significantly affected due to the effluents. Although the values in general were not high, the stretch between Kakinada and Batanagar (polluted zones) appeared to be distinctly affected (Table 5.9). Such affectation sometimes extended both up and down the estuary. The cumulative effect of pollutants present in a sublethal level with longer retention period possibly affected the plankton concentration adversely.

Bottom biota was poorly represented around the outfalls due to concentrated waste-waters containing highly oxidisable organic material, high suspended solids, conservative material, acid, alkali, etc., and also in the main estuary due to scouring effect.

5.4.2 2.2. Pollution from heavy metals : As high as 2.5 mg/l of zinc and 2.0 mg/l of chromium were found in the surface water around the outfall areas of Viscose rayon, rubber and paints and varnish factories discharging 0.1-8.0 mg/l zinc; tannery, paints and varnishes discharging 0.2-2.9 mg/l chromium. The water above the outfall of pulp and paper factory (soda process) was found to contain 0.9 and 0.8 mg/l of zinc and chromium respectively. Their influence was found up to about 3 km downstream, the respective values being 0.8 and 1.2 mg/l at the outfall and 0.7 and 1.1 mg/l at the recovery zone. In sediments, their respective values were 1.1 and 3.5 mg/l. Such higher values were suspected to be brought down from Viscose rayon factory located about 2 km above the paper mill.

TABLE 5.8 **Effect of industrial waste upon the plankton population at the outfall region in the Hooghly estuary in different seasons (P=Phytoplankton, Z=Zooplankton)**

Industry	Location	Season	
		November to February	March to June
(1) Kesoram Rayon	Above the outfall	236 (P—180; Z—56)	195 (P—140; Z—55)
	At the outfall	130 (P— 96; Z—34)	49 (P— 37; Z—12)
	Percentage reduction	42 (P— 46; Z—39)	74 (P— 73; Z—78)
(2) Distillary & Yeast	Above the outfall	89 (P— 48; Z—41)	70 (P— 45; Z—25)
	At the outfall	57 (P— 32; Z—25)	55 (P— 30; Z—25)
	Percentage reduction	36 (P— 36; Z—39)	21 (P— 33; Z— 0)
(3) Tribeni Tissue (Soda process)	Above the outfall	657 (P—555; Z—102)	171 (P—133; Z—38)
	At the outfall	20 (P— 7; Z—13)	87 (P— 54; Z—33)
	Percentage reduction	97 (P— 99; Z—87)	49 (P— 59; Z—23)
(4) India Pulp & Paper (Sulphite process)	Above the outfall	339 (P—259; Z—80)	59 (P— 37; Z—22)
	At the outfall	77 (P— 41; Z—36)	7 (P— 3; Z— 4)
	Percentage reduction	77 (P— 84; Z—55)	88 (P— 92; Z—82)
(5) Rubber	Above the outfall	123 (P— 70; Z—53)	78 (P— 44; Z—34)
	At the outfall	80 (P— 57; Z—23)	55 (P— 40; Z—15)
	Percentage reduction	35 (P— 18.2; Z—56)	26.6 (P— 9; Z—55)

From ; Ray & Laha, 1981.

TABLE 5.9 **Gross and net primary production (mg C/m³/h) of the Hooghly estuary in different seasons (average values only)**

Location	Summer		Monsoon		Winter	
	Gross	Net	Gross	Net	Gross	Net
Kulna	69.0	52.1	37.7	50.2	60.6	54.5
Tribeni	7.3	2.6	31.5	8.3	150.5	92.1
Kasinagar	145.5	58.1	10.4	31.8	56.7	27.1
Kakinada	13.5	0.9	7.0	1.7	23.8	0.0
Barrackpore	46.9	24.7	29.3	8.4	24.7	20.1
Tittaghur	13.8	12.6	5.2	5.2	12.1	12.3
Batanagar	56.2	27.4	11.2	9.8	12.9	14.7
Diamond Harbour	30.9	25.5	25.5	51.6	74.3	51.4
Kakdwip	57.5	15.5	43.0	44.6	79.2	42.2

From ; Ray and Laha, 1981.

The discharge of these metal-borne wastes resulted in bioaccumulation of zinc in shrimp (*Marobrachium* species 0.2-95 ppm), fish (*Aila ooila* 44.8-105.9 ppm; *Setipinna phasa* 0.3-5.6 ppm; *Rita rita* 17.2-37.2 ppm; and *Glossogobius gueris* 8.6-9.6 ppm), while bioaccumulation of chromium in *Setipinna phasa* was between 1.6 and 5.6 ppm (Ghosh and Bagchi—unpublished data). According to Stones (1976), 1 ppm zinc suppresses 8% BOD_5; 1 ppm chromium—18% BOD_5 and 1 ppm copper—44% BOD_5. Further, there are evidences to show that zinc causes decrease of plankton populations. Thus it may be concluded that the polluted stretch of the Hooghly does not provide congenial ecological conditions for the fish populations to prosper.

5.4.2.3. Fish and fisheries

The effect of pollution on fishery was studied by dividing the estuary into two zones with corresponding grouping of fish assembly centres.

Zone I (Unpolluted) : Nabadwip to Bandel near Jubilee

Bridge—A stretch of about 86 km having an approximate water area of 4509.84 ha.

Zone II (Polluted) : Bandel to Baj Baj near Batanagar—A stretch of about 72 km having an approximate water spread area of 4782.24 ha.

Total yield of fish in a period of 16 years covering periods from 1962-63 to 1977-78 at the two zones, have been estimated. Average annual yields in the respective zones were 719.25 and 124.94 tonnes respectively, giving the catch/ha of 159 and 26 kg (Table 5.10). There was also a tendency of increase in yields from year to year with some exceptions. The decrease might have been due to some non-sampling error during the Pre-Farakka barrage period, in the unpolluted zone. On the other hand, polluted zone did not follow a pattern and showed a decreasing trend mostly in 1960's and also in early 1970s, except during the period of bumper catch.

Some recommendations by Ray and Laha (1981) made on the basis of the data quoted above are worth trying for restoration of the dying estuary. They are as follows :

(i) The routine remedial measures taken by the industries and recovery of some basic chemicals were not sufficient to abate pollution, hence they ought to be implemented more vigorously to obtain satisfactory results.

(ii) It is difficult to treat large volumes of wastes, hence more attention ought to be paid to economic recovery of byproducts and to reduce the volume of waste-water.

(iii) Where it was not economical to treat the entire body of waste water, only that particular waste-water which caused pollution, needed to be treated.

(iv) The recycling of such treated waste for other beneficial uses could be thought of.

(v) The discharge of final waste-water, if possible, might be done towards the middle of the river through pipes, during flood tide to keep the littoral zone unaffected.

(vi) Finally, regular periodic monitoring of the estuarine water was essential to ensure safety of water quality, biomass and the estuary in general.

TABLE 5.10 **Total annual landings (in tonnes) in the unpolluted and polluted stretches of the Hooghly estuary**

Year	Unpolluted Zone I	Polluted Zone II
1962-63	562	297
1963-64	484	142
1964-65	338	103
1965-66	536	68
1966-67	720	57
1967-68	742	89
1968-69	735	88
1969-70	910	105
1970-71	913	120
1971-72	1064	120
1972-73	1064	101
1973-74	1029	106
1974-75	779	154
1975-76	681	170
1976-77	551	161
1977-78	400	118
Average	719.25	124.94

From : Ray and Laha, 1981.

CHAPTER VI

Conclusions and Summary

GANGA is one of the oldest rivers to have supported early human settlements in the world. The river has been the place of development of human civilization to a great extent. However, the occupance over the years is now having its ill-effects on the river slowly but surely destroying the river in time and space. The symptoms of degradation have started showing and are getting more severe every day. An urgent need to assess the damage caused by the human settlements and their numerous activities to the great river is being felt. The Planning Commission has, hence, taken steps to introduce efforts for large scale investigations of this river, from origin to the mouth, with the help of a large number of universities and other institutions in the region.

This Report is the first attempt in the country to compile all available information on the river, especially with respect to the human settlements and their activities and their impact on the Ganga river. The Report aims at identifying the areas of concern in different zones of the Ganga.

6.1. HUMAN SETTLEMENTS ALONG THE RIVER

The Ganga Basin has shown an annual increase of 2.5% in its population in the decade 1971-81 with an average density of 464 persons per km^2 in 1981. Urban population of the region showed an annual increase of 7.3% in the same period, reflecting the increasing magnitude of urbanization taking place in the region. The percentage of urban population to total population (18%) however, shows that the region is still mainly rural (Table 6.1).

TABLE 6.1 **Population load in different regions of the Ganga Basin**

Region	Population, 1981		Percentage of urban to total population		Percentage growth rate for 1971-81		Density of population per km² 1981,
	Total	Urban	1971	1981	Total	Urban	
Zone above Upper Ganga Plain	1,805,362	4,35,275	14.49	24.11	+27.18	+97.2	105
Upper Ganga Plain	69,706,811	14,798,000	15.47	21.23	+25.85	+89.71	454
Middle Ganga Plain	83,176,176	8,339,759	7.46	10.03	+24.22	+68.57	593
Lower Ganga Plain	45,242,285	13,285,474	25.45	29.36	+22.08	+36.79	704
Ganga Basin	199,930,634	36,858,508	15.72	18.44	+24.83	+73.07	464

The mode of urbanization is very uneven. It is mainly found in the form of dense pockets in certain regions of the Plains, creating acute problems of population explosion and consequent degradation of the local river quality. There are altogether twenty urban agglomerations situated along or in the vicinity of the Ganga river all along its course. The distribution of these agglomerations in different zones of the great Gangetic Plains is as follows, one in Zone above Upper Ganga Plain, eight in Upper Ganga Plain, seven in Middle Ganga Plain and four in Lower Ganga Plain. Increasing growth of these, have often led to increased uses of the river, many a times, causing degradation of the quality of the river.

The total region is still agrarian in nature. About 70% of the work force is still engaged in pastoral persuits (Table 6.2). However, a net annual shift of 0.34% from agricultural work-force to industrial workforce has taken place in the last decade (Table 6.2), showing the positive increase in industrialization leading to increased urbanization.

6.2. ENVIRONMENTAL PROBLEM CAUSED DUE TO HUMAN SETTLEMENTS, DEVELOPMENTAL ACTIVITIES AND IDENTIFICATION OF AREAS OF CONCERN DIFFERENT ZONES OF THE REGION

The problems are of two kinds—

(i) caused by nature, and (ii) caused by man.

The ill-effects of nature-induced problems are often aggravated by human beings. The problems are discussed below zonewise.

6.2.1. Zone above Upper Ganga Plain

The region is endowed with vast resources of forests, scenic beauty, agricultural and horticultural products and livestock. Such a natural wealth is endangered by the recent increase in human populations and their activities.

Due to hilly terrain and undulating topography the distribution of industries is uneven. At the same time, their

TABLE 6.2 **Working force in different sectors (percentages) as an indicator of human activity in different regions of the Ganga Basin**

Region	Cultivators		Agricultural Labourers		Household industry		Other workers	
	1971	1981	1971	1981	1971	1981	1971	1981
Zone above Upper Ganga Plain	72.94	68.46	2.54	2.36	1.5	1.86	23.01	27.31
Upper Ganga Plain	59.94	59.09	15.74	14.72	3.46	3.74	20.86	22.45
Middle Ganga Plain	44.68	47.21	39.57	34.46	3.12	3.81	12.63	14.69
Lower Ganga Plain	30.26	28.56	29.36	26.57	3.18	4.67	37.20	40.20
Ganga Basin	51.95	50.83	21.80	19.53	2.81	3.52	23.42	26.16

scattering has resulted in no obvious instances of pollution—an offshoot of industrialization.

However, the region faces the peril of urbanization due to an increase in human population (2.7% per year during the last decade) causing stress on the forests especially in the catchments of Bhagirathi, Yamuna and Bhil'angana, which are getting depleted at a faster rate. An increase in the rate of deforestation has created a great impact on the stability of the region. The Ganga has flooded every year during the last decade affecting the regions situated downstream, viz., Upper and Middle Ganga Plain. The floods have caused immense loss of human lives, property, agriculture and livestock, amounting to several crores of rupees. Excessive felling of trees taking place in the region has been the main cause of the 'CHIPCO' movement started by the natives in the area. The movement has gained momentum and worldwide attention for the region.

6.2.2. Upper Ganga Plain

Though the Upper Ganga Plain is relatively more developed, it has its own problems.

The land has been subjected to misuse, leading to several land and associated problems. For instance, over-grazing and deforestation in most areas have led to soil erosion and ravine formations on the one hand and accentuated flooding on the other. The traditional pastoralism of the Braj Plain, a critical rainfall zone, could not maintain the ecosystem partly due to over-grazing and partly due to the encroachment of desertic conditions. Likewise, the vigorous activity in recent years in the use of the Tarai resources, has accentuated the processes of flood, waterlogging and soil erosion.

Due to an increase in exploitation of forests in the zone above the Upper Ganga Plain, there is recurrence of floods in the regions situated downstream. Every year, there is an increase in damage caused by floods, which have led to large-scale destruction of property, crops, public utilities, etc., running into several crores of rupees. A Press report (Times of India, dated 11.10.82) has stated that flood fury in 1982 has cost 183 lives and 57,718 homes in Uttar Pradesh alone.

Besides floods, the region also faces the grave danger of

water pollution especially of the Ganga river, due to increased urbanization though paradoxically, the region itself is still mainly rural. Due to industrial development of cities at an unusually rapid rate, the local pollution problems keep occurring, which sometimes take very acute form. In this region two urban agglomerations, especially Kanpur and Allahabad, are the ones where attention is needed, as their growth and industrial development has been rapid. Kanpur is the largest city of U.P. and is one of the most important industrial towns of northern India. Many industries of different types are situated there and establishment of new ones is also rapid. As a result, the industrial pollution problem is also high. The effluents from industries as well as the untreated domestic waste, enter the stream of river Ganga at several points during its course at Kanpur. The problems of suspended and dissolved solids, oxygen demand and presence of heavy metals are expected to be severe in the region, though hardly any data is available. According to a recent press report (Times of India, dated 9-7-82), a survey by a team of scientists led by Dr. Mohammad Ajmal of the Aligarh Muslim University has revealed high metal contamination of the Ganga at Kanpur, Allahabad and Varanasi. The metals detected were cadmium manganese and cobalt. According to the report, fish samples at Allahabad were also found to have very high levels of metal pollutants.

Allahabad, the other urban agglomeration in the region is comparatively less industrialized as also less urbanized. Even so, the city has not escaped from the pollution problems, especially degeneration of the quality of the Ganga and Yamuna rivers along the banks of which the city is situated. Though the pollution load gets distributed along the two rivers, the problem of pollution does not disappear from the region. The pollution of Ganga due to some industrial complexes is discussed at length in Chapter 3. It would prove educative to examine the different treatment methods for rendering the wastes, industrial or domestic, innocuous before their release into the river at different places.

Another problem of concern is that of mass bathing by millions of pilgrims especially on auspicious days, in the river Ganga. A news item appearing in February 1982 (Times of India, 10-2-82), reflects the severity of the problem. The Central

Board for the Prevention and Control of Water Pollution undertook for the first time, a study during the Ardh Kumbh Mela at Allahabad in January, 1982.

As many as ten million people had a holy dip in the 3 km long bathing stretch on a single day of Mauni Amavasya on January 25th. On another important bathing day, Makar Sankranti on January 14, the number was estimated to be over one million. During the almost one month long Kumbh period, the number of bathers was 50,000 to 60,000 every day against about 4,000 to 5,000 on non-festival days.

The Board, in a preliminary report, expressed that such a large number of people entering a rather limited body of water within a short time exerts pollutional loads on the water stream far beyond its capacity. This increases the danger of water-borne diseases to the bathing community and to the people downstream. The study during the Ardh-Kumbh Mela revealed that the water in the Ganga 10 km upstream of 'Sangam' was satisfactory while 8 km upstream of 'Sangam', the Yamuna waters were good for bathing. The waters at the holy Sangam remained at the satisfactory level on normal days, when there is not too much of mass dipping, but became totally unsatisfactory, especially on crowded days. This persisted even 15 km downstream of the 'Sangam' on the following day of the Kumbh Mela. The study suggested chlorination of the water for bathers' safety.

6.2.3. Middle Ganga Plain

This region is richly fed by rivers. As a consequence, it has to face the ugly side of the well-developed riverine system, i.e., floods and erosion. The land in the region has borne a great burden of these. There are almost no forests, grasslands or pastures or grazing lands, and proper soil management is lacking. As a result, recurrent floods is a permanent feature in the region. Flood damages caused over the last three decades have been mentioned in Chapter 4. In the floods of 1982 (Times of India, dated 11-10 82) Bihar State which comprises the entire Middle Ganga Plain, suffered extensive damage to houses (56,617) and to crops in 12.06 lakh hectares. Extensive afforestation programmes in the catchments of the rivers and all along their course is worth trying if the fury of the floods is to be abated.

The densely populated region also faces the enormous problem of population explosion with all its fallout-shortage of food, malnutrition, diseases, health, education, employment, etc. There is also the problem of diminishing land-holdings or per capita land that gets more acute every day, due to lack of alternative channels of livelihood.

The region, though mainly rural, has shown slow but definite positive trend of urbanization and industrialization, creating problems of pollution. Presently, detection of the problem of pollution is mostly confined to the urban agglomerations and class I cities which are more industrialized but the presence of the problem in smaller cities is also possible. The centres of environmental concern in the area are—Varanasi, Patna, Barauni, Monghyr, Sultanganj, Bhagalpur, Sahebganj, Rajmahal and Farakka. There are very few reports about the quality of Ganga river at these places. Details of the available reports are mentioned in Chapter 4. These show that the industries as well as municipalities of these cities and towns dump their wastes in an untreated form into the river. As a result, the problems of biological pollution and industrial pollution are inevitable, affecting the quality of water to a great extent. The number of bathers is substantially high all along the course of the river in this region also, during festive occasions, retarding the water quality. Cremation of dead bodies and dumping of the ash and partially burnt bodies in the Ganga is a common feature in this region. There is urgent need to discourage these activities in order to retain the physico-chemical, biological as well as aesthetic quality of the river.

6.2.4. Lower Ganga Plain

This region endowed with rich alluvial soil, abundant basic raw materials, numerous perennial streams and a sea-face, attracts a large number of people in search of their livelihood. As a consequence, it is now confronted with certain problems of both natural and man-aided types.

The region, similar to the preceding one upstream, also has to face the brunt of floods from the Ganga and the other rivers. Flood damages in the last three decades have been mentioned in Chapter 5.

The factors which induced rapid economic development in the Hooghly region have themselves generated a chain of complex problems in the very wake of the vigour of development. On the regional level, development was so lop-sided and uncontrolled that even in the exceptionally developed areas (Calcutta), the attendant problems could not be resolved. The most affected area of the lot is the Hooghly estuary which has been the life line of the region for its growth and development so far. Various studies have been carried out along the Hooghly estuary as regards to the impact caused by industrialization on the estuarine eco-system as a whole and some of them have been discussed in Chapter 5. Salient features of these studies stress the needs for immediate attention to the Hooghly estuary for its restoration. Some recommendations made on the basis of the data collected by Ray and Laha (1981), mentioned in Chapter 5, are worth trying for restoration of the dying estuary.

On the whole, the Ganga river system, one of the oldest civilization-supporting system in the world, requires immediate attention, if it is to continue to be the holy river of the masses. The huge quantum of water flowing in the river, is capable of diluting most of the impurities added to it according to some scientists who claim to assign some exceptional self-purifying properties to the river (Bhargava, 1981b). Such a claim however, may contribute to the perpetuation of complacency on the part of the people who should know better. It is, as a matter of fact, an indication of how little we know about the river we revere most. In spite of a large number of educational and other institutions, both old and new, located along the length of the river, very little data on the quality of the river has been generated to arrive at definite conclusions regarding the accurate status of the quality of water. Number of human settlements, density of human populations and variety of human occupations stress the need for generation of this data on a large scale and on taking corrective measures, ensuring that the river continues to be the life-line of large body of humanity.

Suggested Plan of Action

1. ANY activity leading to deforestation has to be discouraged at the earliest possible time. Afforestation and establishment of vegetation in the watershed as well as on the banks all along the course of the river should be undertaken. This can be undertaken by the forest department of States of U.P., Bihar and West Bengal. Special attention is needed in the catchment area of the Ganga in the zone above Upper Ganga Plain in the U.P. State, where deforestation is on an increase in recent times.

2. Problem of mass bathing on auspicious days, in the Ganga is causing grave concern as the quality of water gets deteriorated as observed from recent reports. A need to discourage this activity is felt urgently. Authorities from States' concern should enforce some measures to prevent water pollution on this score, by introducing measures for improved hygiene at the pilgrimage centres. This would prevent outbreaks of epidemics and safety of the bathers as well as of permanent residents of the regions would be ensured.

3. An understanding of the nutrient budgets and mineral flows in the areas lying along the banks or in the vicinity of the river, has to be developed for evaluation of the agricultural productivity of different regions along the river. This should be done by selected institutions with expertise available to them. Reference in these studies should be made to floods and soil erosion of routine as well as episodal nature.

4. Treatment of waste-water from towns and cities situated all along the course of the river has to be undertaken by the locally concerned authorities like corporations, municipalities, etc.

5. Strict enforcement of laws for the treatment of waste-water from industries situated all along the course of the river for maintaining quality of water needs to be undertaken.

6. Regular monitoring of water quality at different centres like Hardwar, Kanpur, Allahabad, Varanasi, Patna, Barauni, Monghyr, Bhagalpur, Calcutta and Howrah should be planned by Central and State bodies.

7. There are claims that the river has great self-purifying capacities. Detailed scientific investigations using known expertise in the field have to be undertaken in large number of areas along the river body to identify and quantify the capacity. Measures should be taken to derive suitable benefits from the unique property of the river.

References

Bhargava, D.S. (1981a). Pollution Control Strategy for Ganga, In : *Proceedings of International Symposium on Water Resources Conservation, Pollution and Abatement*, (ed.) Uday Chitranshi, Roorkee, December 11-13, 1981, pp 31-38.

Bhargava, D.S. (1981b). Ganga—The most self-purifying river, In : *Proceedings of International Symposium on Water Resources Conservation, Pollution and Abatement*, (ed.) Uday Chitranshi, Roorkee, December 11-13, 1981.

Bilgrami, K.S. and Datta Munshi, J.S. (1979) Limnological Survey and Impact of Human Activities on the river Ganges (Barauni to Farakka range), Technical report prepared under Man & Biosphere Programme, Project 5 UNESCO, at P.G. Departments of Botany Zoology, Bhagalpur University, Bhagalpur, during June, 1976 May, 1979.

Census of India, (1981). Series 1—India, Paper 2 of 1981, Provisional population totals : Rural-Urban Distribution, compiled by P. Padmanabha of the Indian Administrative Service, Registrar General and Census Commissioner for India, printed by the Manager, Govt. of India Press, New Delhi.

Census of India, (1981). Series 1—India, Paper 3 of 1981, Provisional population totals : Workers and Non-workers, compiled by P. Padmanabha of the Indian Administrative Service, Registrar General and Census Commissioner for India, printed by the Manager, Govt. of India Press, Nasik.

Chandra, K.; Singh, B.; Srivastava, G.N. and Mehrotra, S.N. (1981). Pollution from wastes of industries manufacturing nitrogenous fertilizer—a case study from river Ganga near Allahabad (India), In : *Proceedings of International Symposium on Water Resources Conservation, Pollution and Abatement*, (ed.) Uday Chitranshi, Roorkee, December 11-13, 1981, pp. 141-51.

Chaphekar, S.B. (co-ordinator) (1979). Studies of the Kalu river ecosystem—A report on an interdisciplinary project carried out at the Institute of Science, Bombay-32, January-June, 1978.

Das, R.K. (1968). Four Types of Agricultural Lands in Bankura, *Geographical Review of India*, 23 : 2.

Department of Agriculture, U.P. (1976). *Bhagirathi, Bhillangana and*

Yamuna Catchments, Vol. I Inventory Results, Pre-Investment Survey of Forest Resources, Dehra Dun.

Food and Agriculture Organisation (1967). Water Pollution Research with respect to Inland Fisheries, Report of the FAO/UNDP (TA) N 2449, p. 76.

Ghosh, B.B.; Ray, P. and Gopalakrishnan, V. (1973). Survey and characterisation of waste water discharged into the Hooghly estuary, India. *Journal of the Inland Fisheries Society of India*, 5.

Ghosh, B. (1977). *An Approach to Flood Problem of Ganga Basin*, Central Board of Irrigation & Power (CBIP), Golden Jubilee Commemorative volume (1927-77) pp. 73-77, published by C.V.J. Verma, Secretary, CBIP, New Delhi.

Gopalakrishnan, V. (1968). Fishery resources of the Hooghly-Matlah estuarine system and its relation to fisheries of Bay of Bengal. In : *Symposium on Living Resources of the Seas Around India* (From Jhingran, 1977).

Jhingran, V.G. (1977). *Fish and Fisheries of India*. Hindustan Publishing Corporation (India), New Delhi.

Jhingran, V.G. *et al.*, (1970). *Fisheries of the Ganga River System*, (MS) (From Jhingran, 1977).

Kar, N.R. (1968). Calcutta Conurbation. In : *India-Regional Studies* (ed.) Singh, R.L. (Indian National Committee for Geography, p. 337.

Kaushic, S.D. (1962). Agriculture in the Himalayas, *Nat. Geog. Jour., India*, 8: 270-83.

Kaushic, S.D. (1965). Hydroelectric power potential in the Himalayan districts of U.P., Himachal and Punjab, Agra University. *Jour. of Res.* (Letters), 13 : 75-92.

Mathur, R.N. (1961). Some characteristic features of water table in Meerut district, U.P., *N.G.J.I.*, 8-269.

Meher Homji, V.M.; Fontanel, J.; Pascal, J.P.; Chandrahasan, K.; Delacourt, A. and Legris, P. (1978). Notes on the sheet—Allahabad, Extrait des Travaux de la Section Scientifique of Technique de L'Institut Francais de Pondichery, Hors Serie No. 15.

Mitra, G.N. (1967). Development of fisheries in the Sunderbans and the Bay of Bengal (Govt. of India, Project Report I), 63 p. (From Jhingran, 1977).

NEERI (1978). Water quality survey of the Hooghly estuary (1974-76), sponsored by Calcutta Metropolitan Development Authority, Calcutta. (From Ray & Laha, 1981).

Niyogi, D. and Chakravarti, A. (1967). Applied Geomorphology along Digha Beach, Midnapur District, West Bengal, In : *Proceedings of Geomorphological Studies in India*, held at Saugar University, 1967, p 205.

Pant, S.D. (1935). *The Social Economy of the Himalayans*. George Allen & Unwin Ltd., London, 59.

Planning Commission (Education Division) (1981). Conference of the Vice-Chancellors and Directors of Research Institutes for promoting a co-ordinated study of the Ganga in all its aspects. U.P. No. Q-17022 6/81-Edn. dated 11-12-1981.

Puri, G.S. (1960). *Indian Forest Ecology*, Oxford Book & Stationery Co., New Delhi, 548.

Rai, L.C. (1978). Ecological studies of algal communities of the Ganges river at Varanasi, *Indian J. Ecol.*, 5(1) : 1-6.

Rao, K.L. (1975). *India's Water Wealth—Its Assessment, Uses and Projections.* Published by Surjit Mukherji for Orient Longman Ltd., New Delhi.

Ray, P. (1980). Impact of man on the Hooghly estuary, A comprehensive review under pollution management, prepared for the School of Environmental Sciences, J.N.V. (From Ray & Laha, 1981).

Ray, P. and Laha, G.C. (1981). Effects on industrial and municipal wastes on the trend of fisheries in the Hooghly estuary, In : *Proceedings of the International Symposium on Water Resources Conservation, Pollution and Abatment* (ed.) Uday Chitranshi, Roorkee, December 11-13, 1981, pp. 121-130.

Register of Large Dams in India (1979). Published by Indian National Committee for ICOLD, Central Board of Irrigation and Power, New Delhi.

Saxena, K.L.; Chakraborty, R.N.; Khan, A.Q.; Chattopadhyaya, S.N. and Chandra, H. (1966). Pollution studies of the river Ganges near Kanpur, *Environ. Hlth.*, 8(4) : 270-285.

Sen, Satyakam (1970). The importance of drainage in agriculture of West Bengal. In : *West Bengal* (eds) A.B. Chatterjee *et al.*, Geographical Institute, Presidency College, Calcutta, p. 79.

Singh, S.C. (1965a). Delimitation of the Middle Ganga Plain, *N.G.J.I.*, 11 : 74-83.

Singh, S.C. (1965b). Changes in the courses of rivers and their effects on urban settlements in the Middle Ganga Plain, Ph. D. Thesis, Deptt. of Geography, B.H.U., Varanasi.

Singh, R.L. (ed.) (1971). *India—A Regional Geography* Silver Jubilee Publication of National Geographical Society of India, Varanasi.

Spate, O.H.K. and Learmonth, A.T.A. (1967). India and Pakistan. A General and Regional Geography, Methuen & Co. Ltd., p. 571.

Stones, T. (1976). A discourse on the biochemical oxidation of sewage, *Effluent Water Treatment Journal*, 16 : 352-355.

Index